W9-CAH-244

student study
ART NOTEBOOK

Human BIOLOGY
fourth edition

Sylvia S. Mader

WCB **Wm. C. Brown Publishers**

Dubuque, IA Bogota Boston Buenos Aires Caracas Chicago
Guilford, CT London Madrid Mexico City Sydney Toronto

Wm. C. Brown Communications, Inc.

President and Chief Executive Officer *G. Franklin Lewis*
Corporate Senior Vice President, President of WCB Manufacturing *Roger Meyer*
Corporate Senior Vice President and Chief Financial Officer *Robert Chesterman*

The credits section for this book begins on page 147 and
is considered an extension of the copyright page.

Printed in the United States of America by Wm. C. Brown Communications, Inc.,
2460 Kerper Boulevard, Dubuque, IA 52001

10 9 8 7 6 5 4 3 2 1

TO INSTRUCTORS AND STUDENTS

This Student Study Art Notebook is free with a new textbook to all students and can be used to take notes during lectures. On each notebook page, there are two figures (sometimes one, sometimes three) faithfully reproduced from the original textbook figure. Each figure also corresponds to each of the 200 acetates available to instructors with adoption of the text.

The intention is to place a copy of the transparency acetate art in front of students (via the notebook) as the instructor uses the overhead during lectures. The advantage to the student is that he/she will be able to see all labels clearly, and take meaningful notes without having to make hurried sketches of the acetate figure.

The pages of the Art Notebook are perforated and three-hole punched, so they can be removed and placed in a personal binder for specific study and review, or to create space for additional notes.

DIRECTORY OF NOTEBOOK FIGURES

TO ACCOMPANY

HUMAN BIOLOGY, 4TH ED. BY

SYLVIA S. MADER

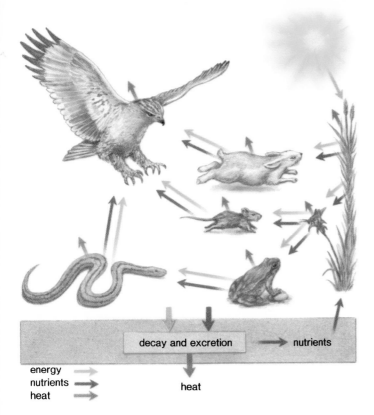

Chemical Cycling and Energy Flow
Figure I.2

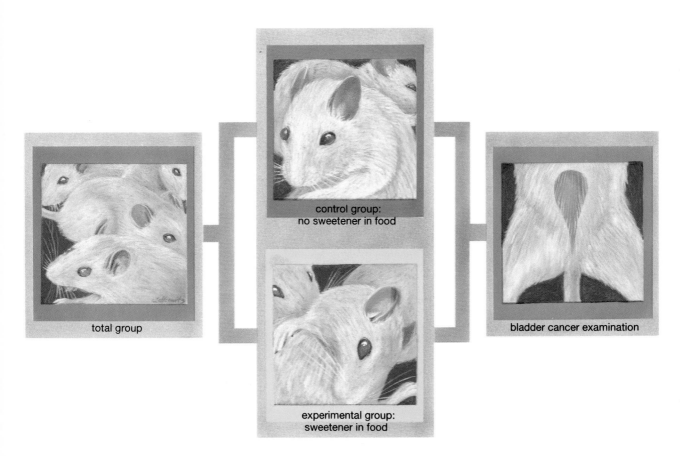

Controlled Experiment
Figure I.5

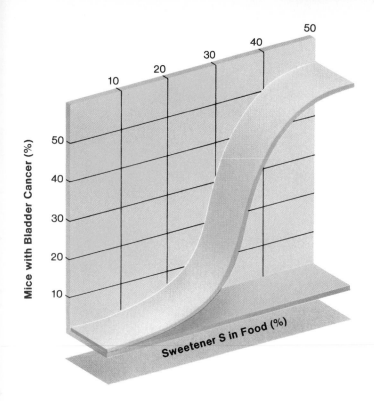

Presenting the Data
Figure I.6

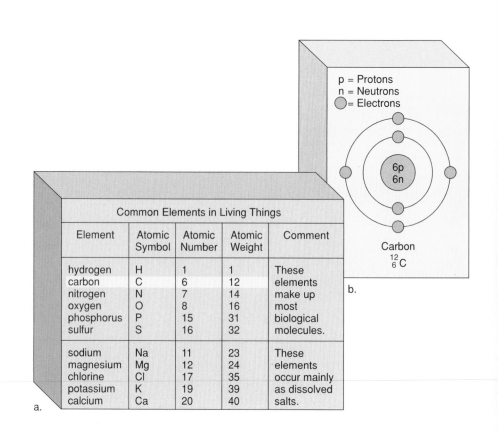

Common Elements in Living Things				
Element	Atomic Symbol	Atomic Number	Atomic Weight	Comment
hydrogen	H	1	1	These
carbon	C	6	12	elements
nitrogen	N	7	14	make up
oxygen	O	8	16	most
phosphorus	P	15	31	biological
sulfur	S	16	32	molecules.
sodium	Na	11	23	These
magnesium	Mg	12	24	elements
chlorine	Cl	17	35	occur mainly
potassium	K	19	39	as dissolved
calcium	Ca	20	40	salts.

a.

p = Protons
n = Neutrons
◯ = Electrons

6p
6n

Carbon
$^{12}_{6}C$

b.

Elements and Atoms
Figure 1.1

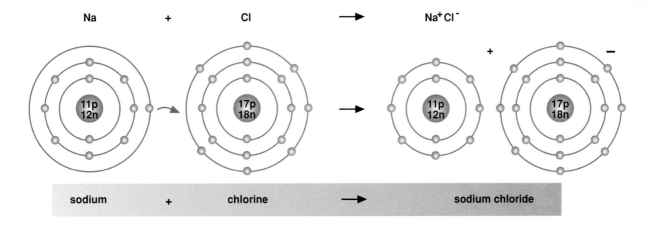

Formation of Sodium Chloride
Figure 1.3

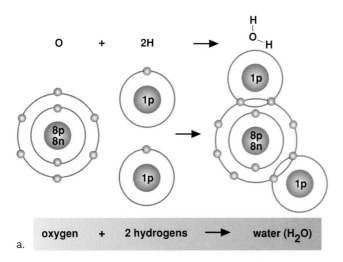

Formation of Water
Figure 1.4 a

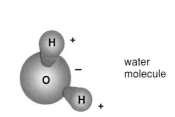

3-D Model of Water Molecule
Figure 1.4 b

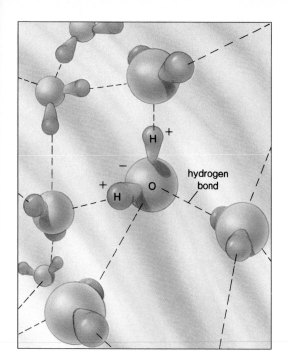

Hydrogen Bonding Between Water
Figure 1.5

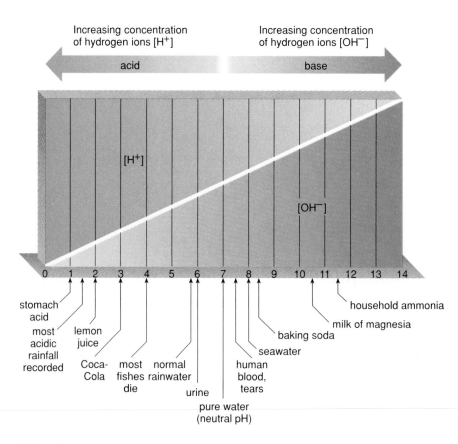

Increasing concentration
of hydrogen ions [H⁺]

Increasing concentration
of hydrogen ions [OH⁻]

acid

base

$[H^+]$

$[OH^-]$

0 1 2 3 4 5 6 7 8 9 10 11 12 13 14

stomach
acid

most
acidic
rainfall
recorded

most
fishes
die

normal
rainwater

urine

human
blood,
tears

pure water
(neutral pH)

seawater

baking soda

milk of magnesia

household ammonia

lemon
juice

Coca-
Cola

pH Scale
Figure 1.6

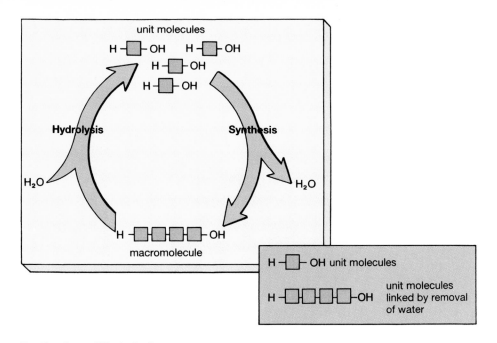

Synthesis and Hydrolysis
Figure 1.7

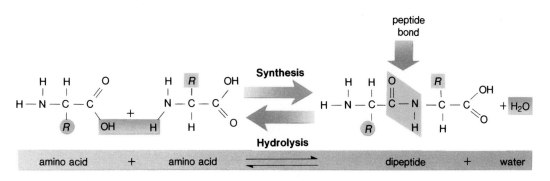

Formation of a Dipeptide
Figure 1.9

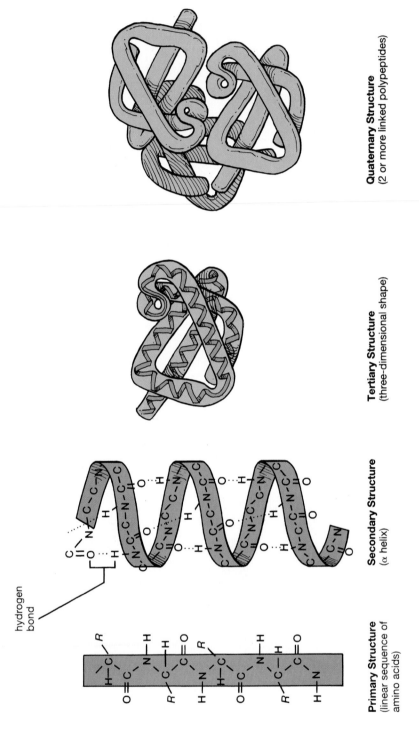

Primary Structure
(linear sequence of
amino acids)

hydrogen
bond

Secondary Structure
(α helix)

Tertiary Structure
(three-dimensional shape)

Quaternary Structure
(2 or more linked polypeptides)

Levels of Protein Structure
Figure 1.10

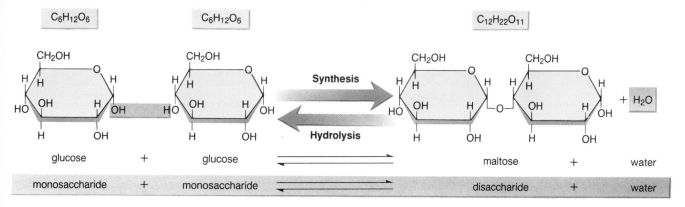

Synthesis and Hydrolysis of Maltose
Figure 1.11

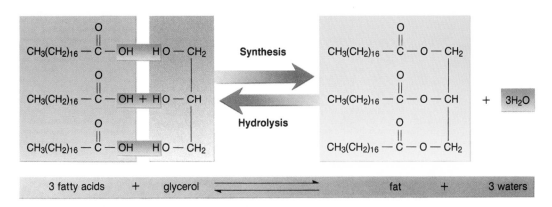

Synthesis and Hydrolysis of Fat
Figure 1.14

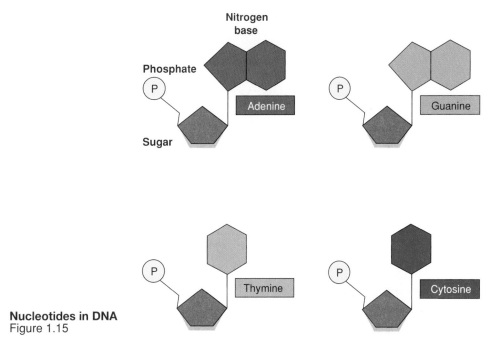

Nucleotides in DNA
Figure 1.15

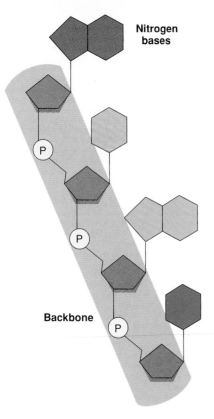

Nitrogen
bases

Backbone

Nucleic Acid Strand
Figure 1.16

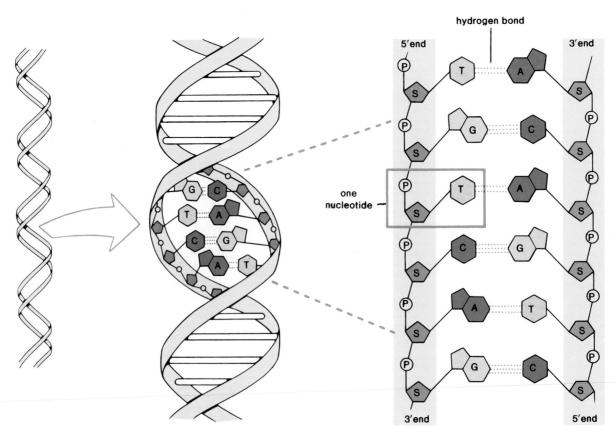

hydrogen bond

5'end

3'end

one
nucleotide

3'end

5'end

Overview of DNA Structure
Figure 1.17

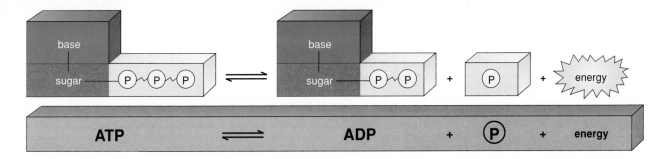

ATP Reaction
Figure 1.18

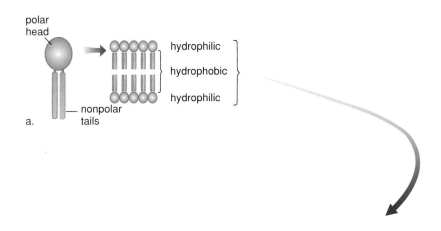

a.

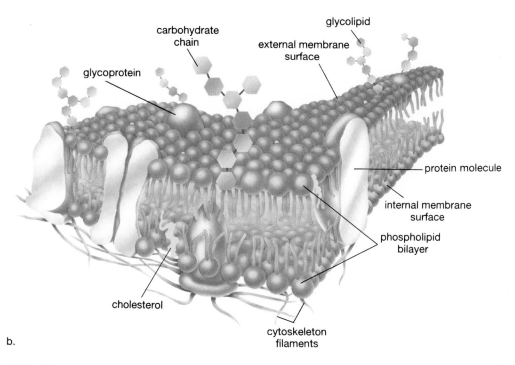

b.

Cell Membrane
Figure 2.2

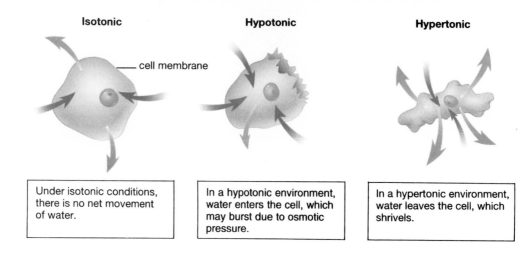

Isotonic

cell membrane

Hypotonic

Hypertonic

Under isotonic conditions, there is no net movement of water.

In a hypotonic environment, water enters the cell, which may burst due to osmotic pressure.

In a hypertonic environment, water leaves the cell, which shrivels.

Tonicity Diagram
Figure 2.3

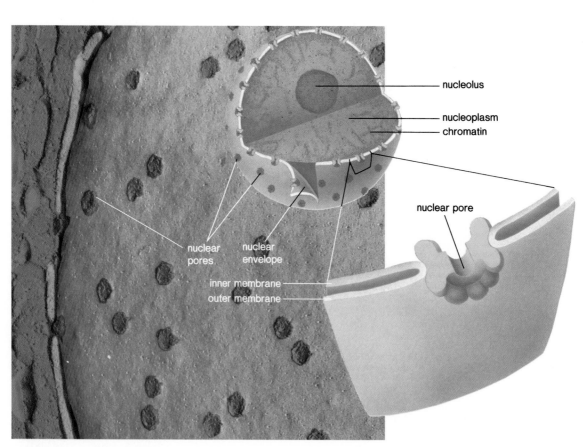

nucleolus

nucleoplasm

chromatin

nuclear pore

nuclear pores

nuclear envelope

inner membrane

outer membrane

Anatomy of the Nucleus
Figure 2.5

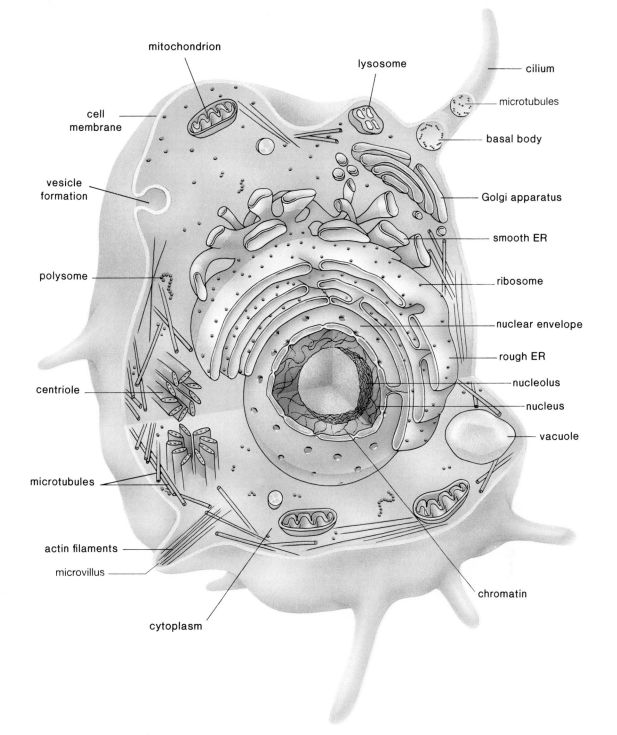

mitochondrion

lysosome

cilium

microtubules

basal body

cell
membrane

Golgi apparatus

vesicle
formation

smooth ER

ribosome

polysome

nuclear envelope

rough ER

nucleolus

centriole

nucleus

vacuole

microtubules

actin filaments

microvillus

chromatin

cytoplasm

Animal Cell
Figure 2.4

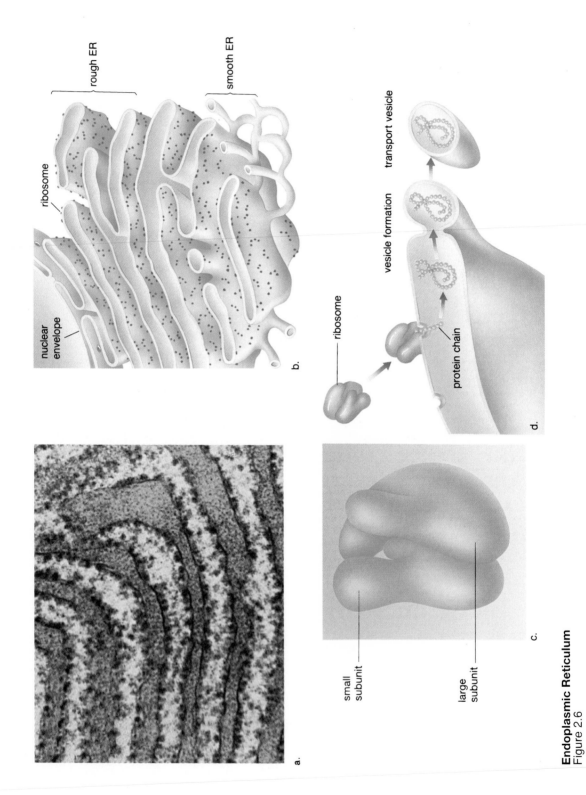

Endoplasmic Reticulum
Figure 2.6

a.

b.

rough ER

smooth ER

ribosome

nuclear
envelope

c.

small
subunit

large
subunit

d.

ribosome

transport vesicle

vesicle formation

protein chain

12

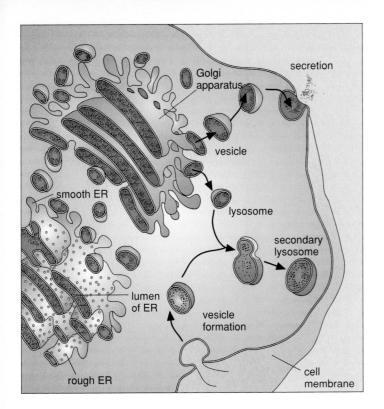

Golgi Apparatus Structure
Figure 2.7

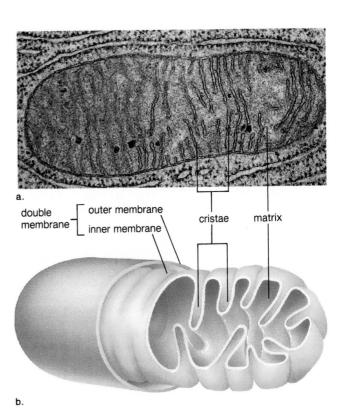

a.

double membrane ⎡ outer membrane ⎣ inner membrane   cristae   matrix

b.

Mitochondrion Structure
Figure 2.8

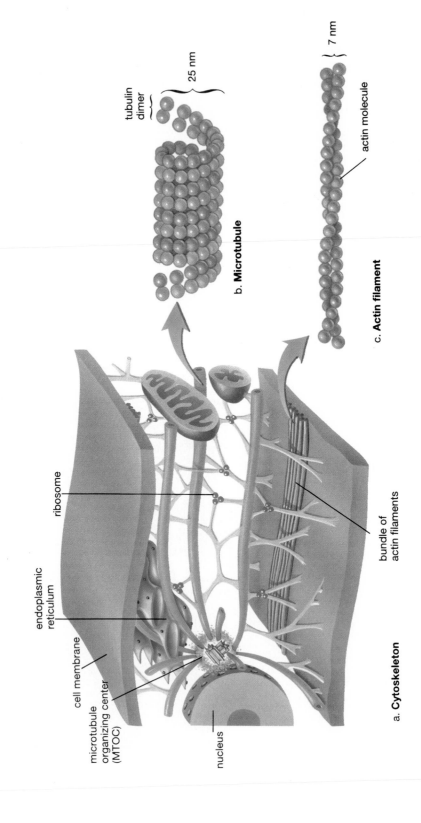

cell membrane

microtubule
organizing center
(MTOC)

endoplasmic
reticulum

ribosome

nucleus

bundle of
actin filaments

a. **Cytoskeleton**

tubulin
dimer

25 nm

b. **Microtubule**

7 nm

actin molecule

c. **Actin filament**

The Cytoskeleton
Figure 2.9

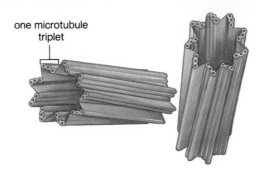

Centrioles
Figure 2.10

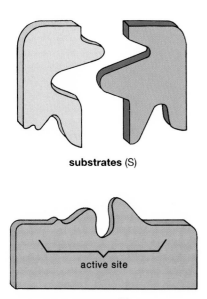

substrates (S)

active site

enzyme (E)

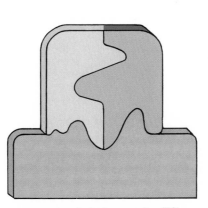

enzyme-substrate complex (ES)

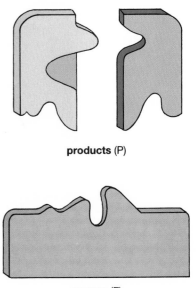

products (P)

enzyme (E)

Enzymatic Action
Figure 2.12

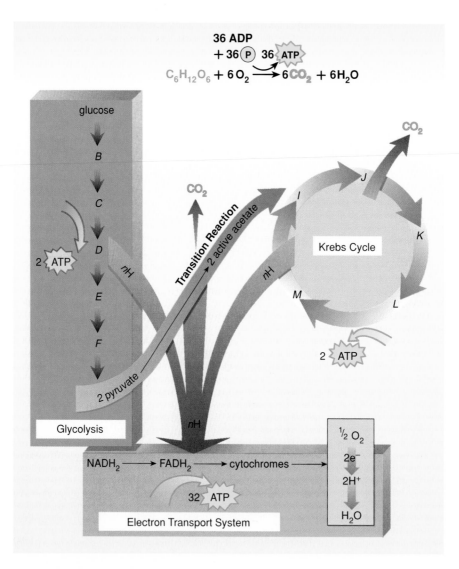

Aerobic Cellular Respiration
Figure 2.13

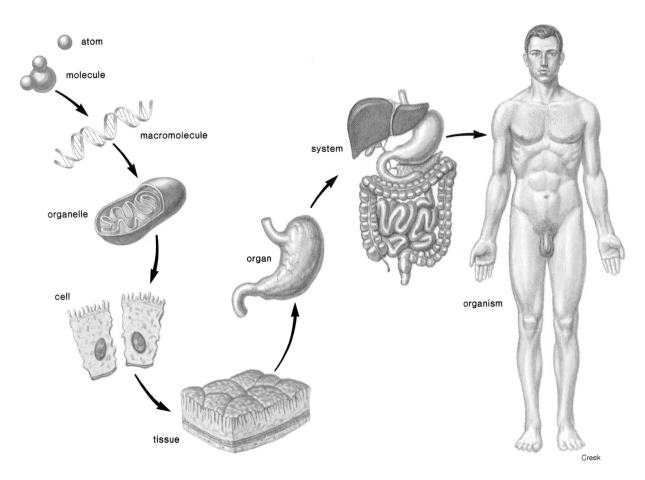

atom

molecule

macromolecule

organelle

cell

tissue

organ

system

organism

Creek

Levels of Organization
Figure 3.1

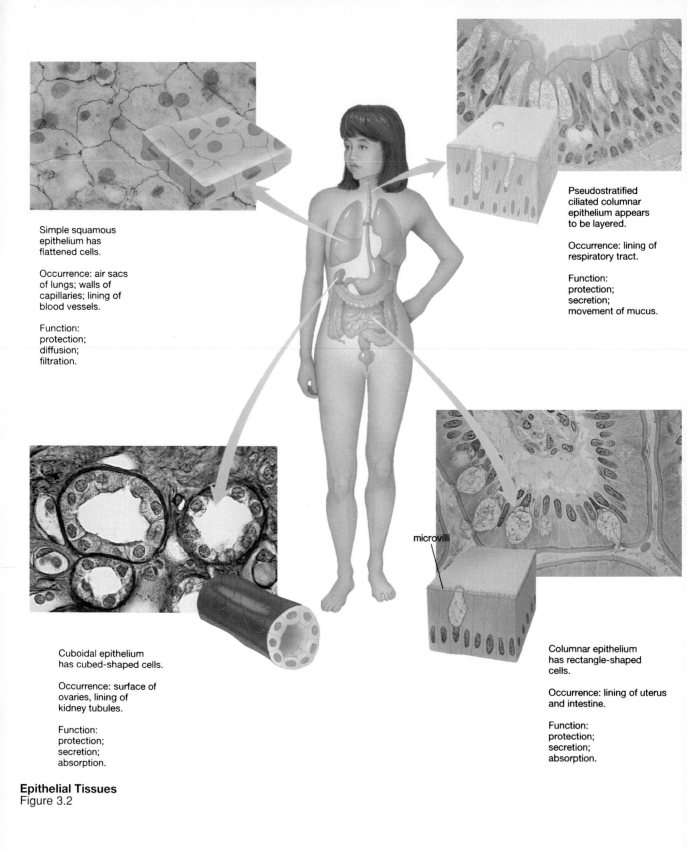

Simple squamous epithelium has flattened cells.

Occurrence: air sacs of lungs; walls of capillaries; lining of blood vessels.

Function: protection; diffusion; filtration.

Pseudostratified ciliated columnar epithelium appears to be layered.

Occurrence: lining of respiratory tract.

Function: protection; secretion; movement of mucus.

Cuboidal epithelium has cubed-shaped cells.

Occurrence: surface of ovaries, lining of kidney tubules.

Function: protection; secretion; absorption.

microvilli

Columnar epithelium has rectangle-shaped cells.

Occurrence: lining of uterus and intestine.

Function: protection; secretion; absorption.

Epithelial Tissues
Figure 3.2

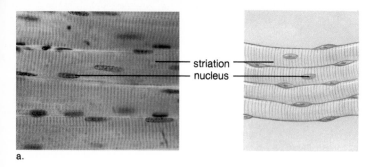

a.

Skeletal Muscle
Figure 3.6 a

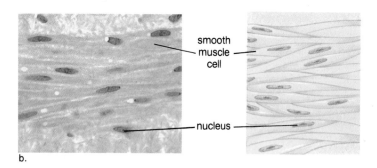

b.

Smooth Muscle
Figure 3.6 b

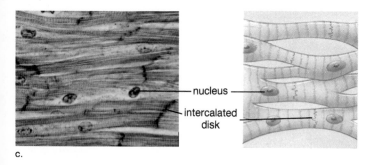

c.

Cardiac Muscle
Figure 3.6 c

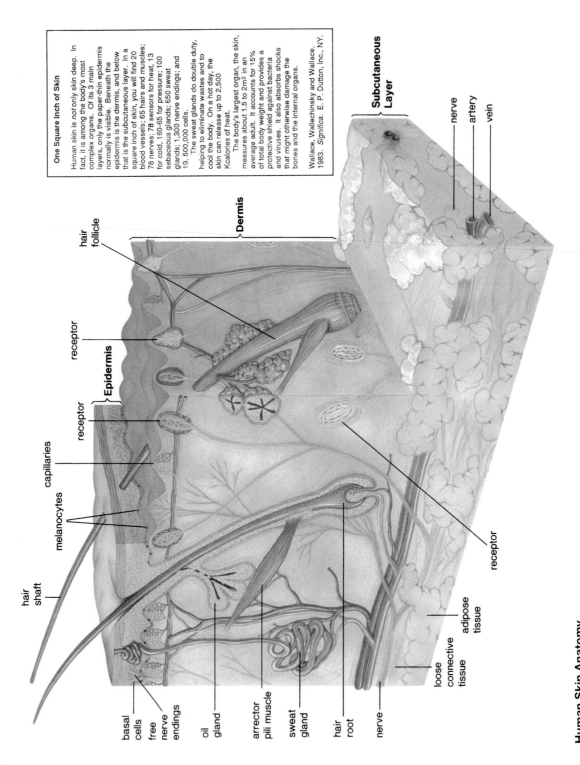

One Square Inch of Skin

Human skin is *not* only skin deep. In fact, it is among the body's most complex organs. Of its 3 main layers, only the paper-thin epidermis normally is visible. Beneath the epidermis is the dermis, and below that is the subcutaneous layer. In a square inch of skin, you will find 20 blood vessels; 65 hairs and muscles; 78 nerves; 78 sensors for heat, 13 for cold, 160-65 for pressure; 100 sebaceous glands; 650 sweat glands; 1,300 nerve endings; and 19, 500,000 cells.

The sweat glands do double duty, helping to eliminate wastes and to cool the body. On a hot day, the skin can release up to 2,500 Kcalories of heat.

The body's largest organ, the skin, measures about 1.5 to 2m² in an average adult. It accounts for 15% of total body weight and provides a protective shield against bacteria and viruses. It also absorbs shocks that might otherwise damage the bones and the internal organs.

Wallace, Wallechinsky and Wallace. 1983. *Significa.* E. P. Dutton, Inc., NY.

Dermis

Epidermis

Subcutaneous Layer

hair follicle

receptor

receptor

capillaries

melanocytes

hair shaft

basal cells

free nerve endings

oil gland

arrector pili muscle

sweat gland

hair root

nerve

receptor

nerve

artery

vein

adipose tissue

loose connective tissue

Human Skin Anatomy
Figure 3.8

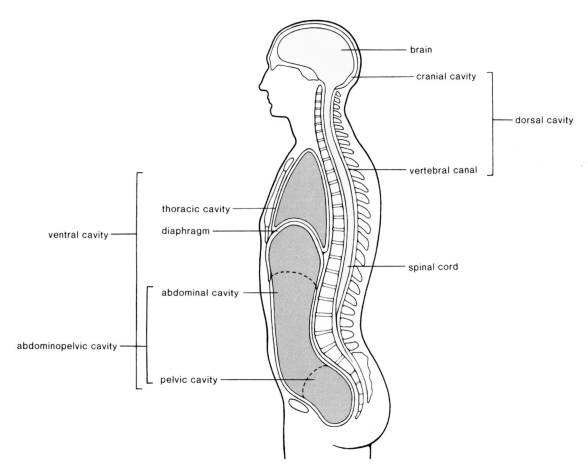

Organization of the Human Body
Figure 3.9

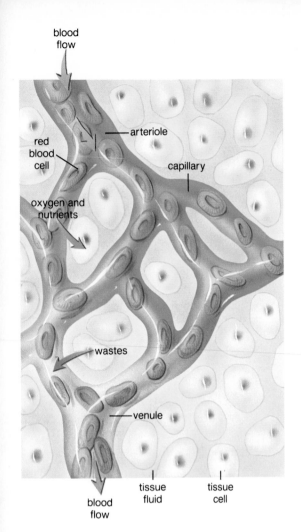

Formation of Tissue Fluid
Figure 3.10

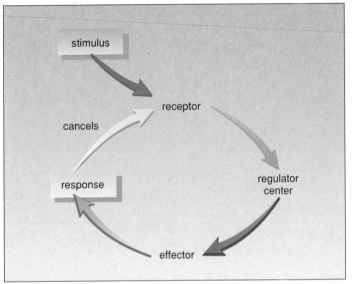

a.

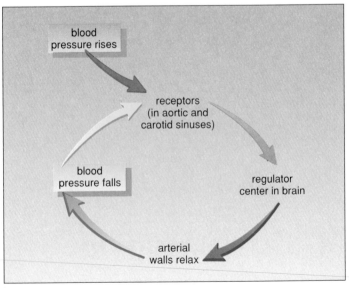

b.

Negative Feedback Control
Figure 3.12

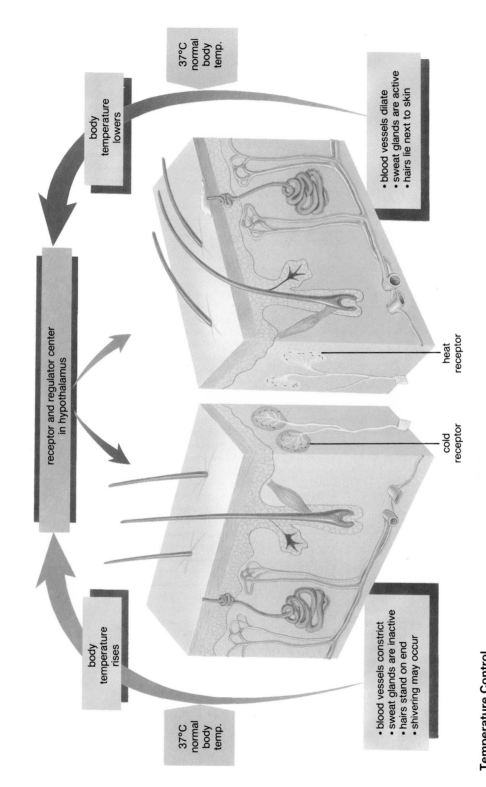

37°C normal body temp.

body temperature lowers

receptor and regulator center in hypothalamus

- blood vessels dilate
- sweat glands are active
- hairs lie next to skin

heat receptor

cold receptor

body temperature rises

37°C normal body temp.

- blood vessels constrict
- sweat glands are inactive
- hairs stand on end
- shivering may occur

Temperature Control
Figure 3.13

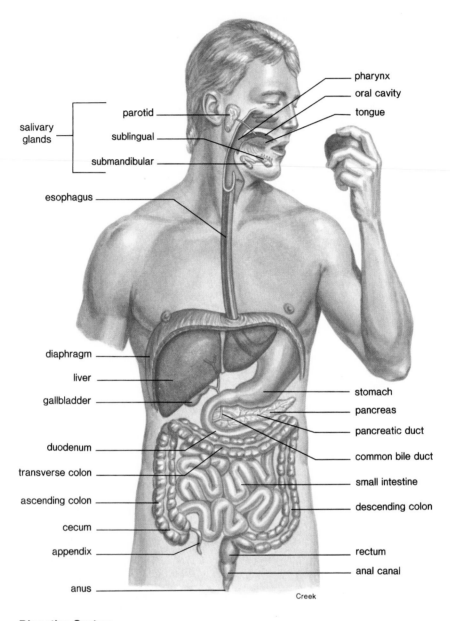

pharynx
oral cavity
tongue
salivary glands
parotid
sublingual
submandibular
esophagus
diaphragm
liver
gallbladder
duodenum
transverse colon
ascending colon
cecum
appendix
anus
stomach
pancreas
pancreatic duct
common bile duct
small intestine
descending colon
rectum
anal canal

Creek

Digestive System
Figure 4.1

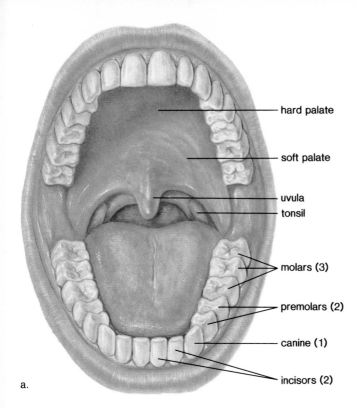

hard palate

soft palate

uvula

tonsil

molars (3)

premolars (2)

canine (1)

incisors (2)

a.

Diagram of Mouth
Figure 4.2 a

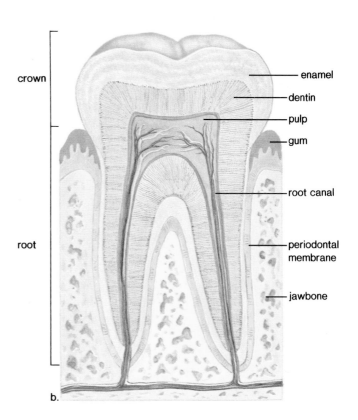

crown

root

enamel

dentin

pulp

gum

root canal

periodontal
membrane

jawbone

b.

Diagram of Tooth
Figure 4.2 b

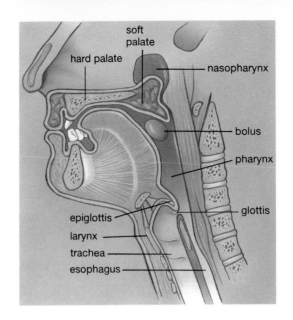

Swallowing
Figure 4.3

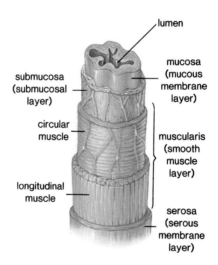

Diagram of Esophagus
Figure 4.4

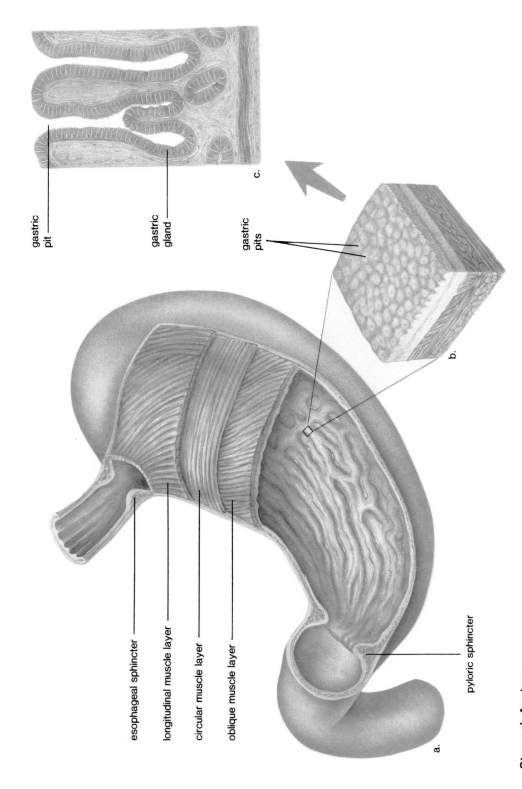

gastric
pit

gastric
gland

gastric
pits

c.

b.

esophageal sphincter

longitudinal muscle layer

circular muscle layer

oblique muscle layer

pyloric sphincter

a.

Stomach Anatomy
Figure 4.5

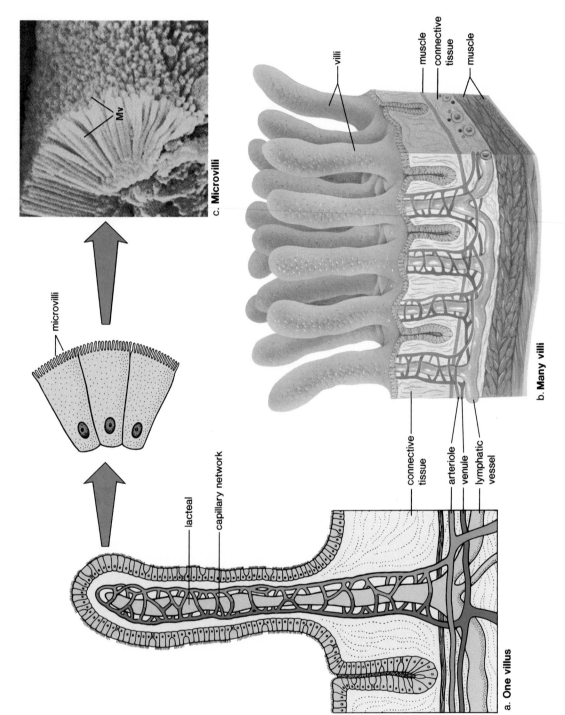

c. **Microvilli**

Mv

microvilli

vilii

muscle

connective tissue

muscle

b. **Many villi**

lacteal

capillary network

connective tissue

arteriole

venule

lymphatic vessel

a. **One villus**

Intestinal Villi
Figure 4.6

28

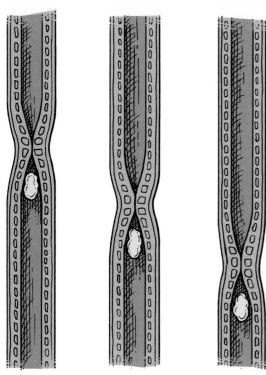

Peristalsis
Figure 4.8

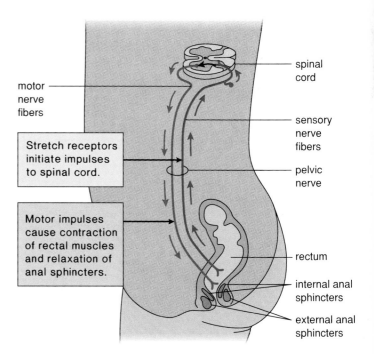

motor
nerve
fibers

spinal
cord

sensory
nerve
fibers

pelvic
nerve

Stretch receptors
initiate impulses
to spinal cord.

Motor impulses
cause contraction
of rectal muscles
and relaxation of
anal sphincters.

rectum

internal anal
sphincters

external anal
sphincters

Defecation Reflex
Figure 4.9

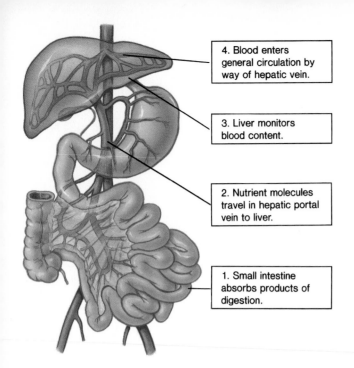

4. Blood enters general circulation by way of hepatic vein.

3. Liver monitors blood content.

2. Nutrient molecules travel in hepatic portal vein to liver.

1. Small intestine absorbs products of digestion.

Hepatic Portal System
Figure 4.10

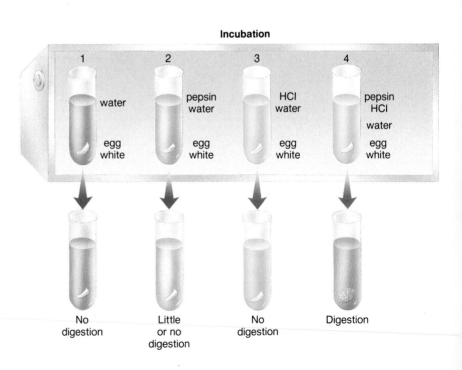

Incubation

1	2	3	4
water	pepsin water	HCl water	pepsin HCl water
egg white	egg white	egg white	egg white

No digestion | Little or no digestion | No digestion | Digestion

Digestive Enzyme Experiment
Figure 4.12

30

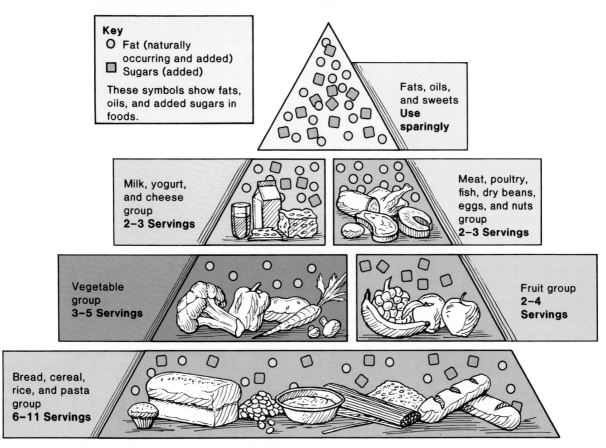

Key
○ Fat (naturally occurring and added)
□ Sugars (added)
These symbols show fats, oils, and added sugars in foods.

Fats, oils, and sweets
Use sparingly

Milk, yogurt, and cheese group
2–3 Servings

Meat, poultry, fish, dry beans, eggs, and nuts group
2–3 Servings

Vegetable group
3–5 Servings

Fruit group
2–4 Servings

Bread, cereal, rice, and pasta group
6–11 Servings

Food Guide Pyramid: A Guide to Daily Food Choices

Food Pyramid
Figure 4.13

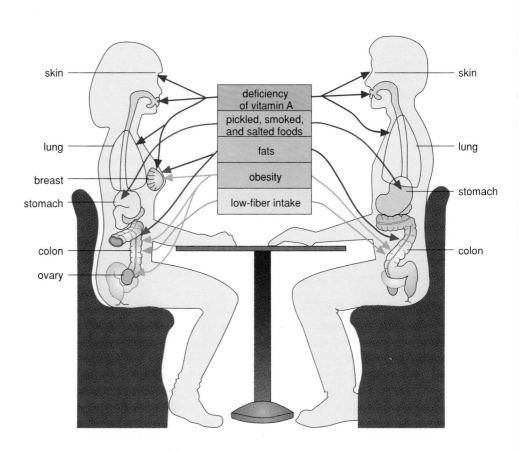

skin

lung

breast

stomach

colon

ovary

deficiency of vitamin A
pickled, smoked, and salted foods
fats
obesity
low-fiber intake

skin

lung

stomach

colon

Diet and Cancer
Figure 4.15

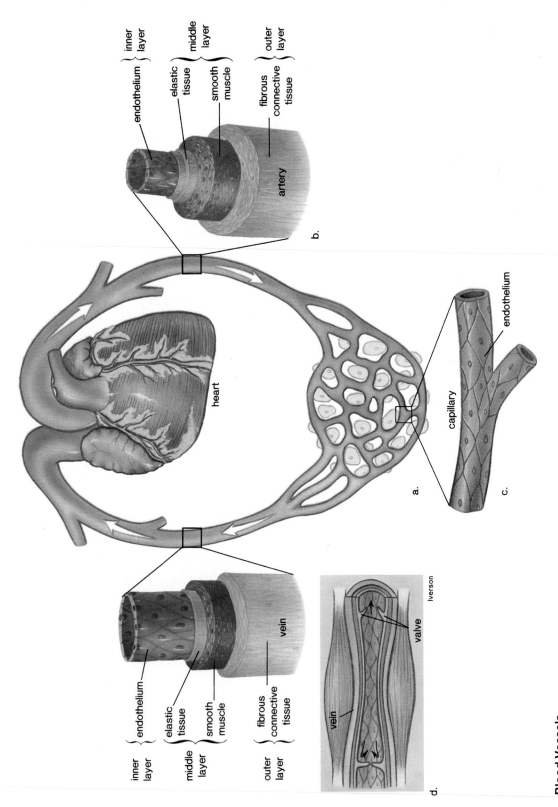

inner layer
- endothelium

middle layer
- elastic tissue
- smooth muscle

outer layer
- fibrous connective tissue

artery

b.

heart

a.

capillary

endothelium

c.

inner layer
- endothelium

middle layer
- elastic tissue
- smooth muscle

outer layer
- fibrous connective tissue

vein

vein

valve

Iverson

d.

Blood Vessels
Figure 5.1

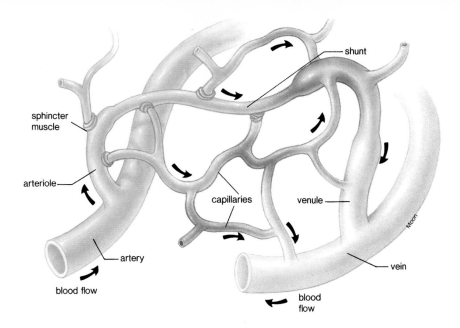

Anatomy of a Capillary Bed
Figure 5.2

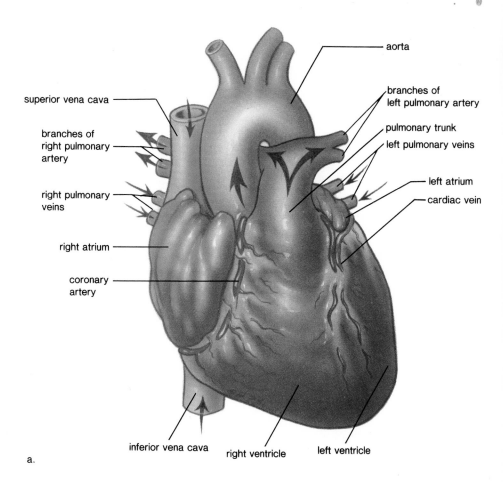

a.

Surface View of the Heart
Figure 5.3 a

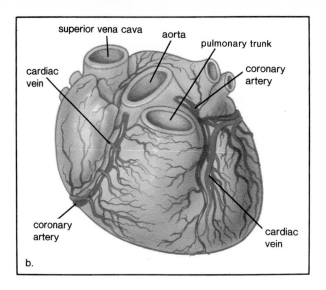

Coronary Arteries and Cardiac Veins
Figure 5.3 b

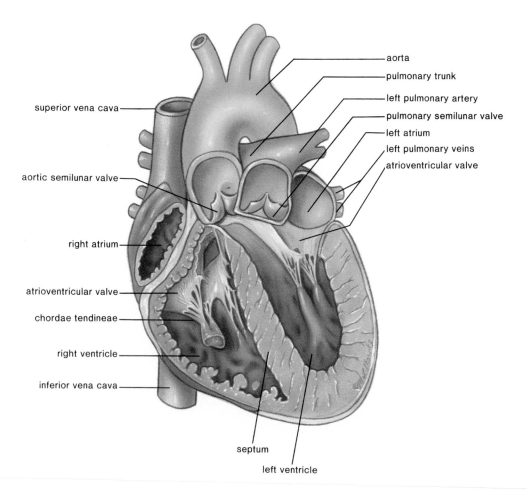

Internal View of the Heart
Figure 5.4 a

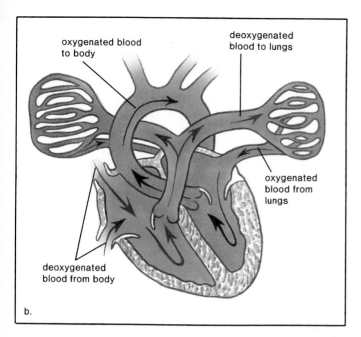

oxygenated blood
to body

deoxygenated
blood to lungs

oxygenated
blood from
lungs

deoxygenated
blood from body

b.

Path of Blood through the Heart
Figure 5.4 b

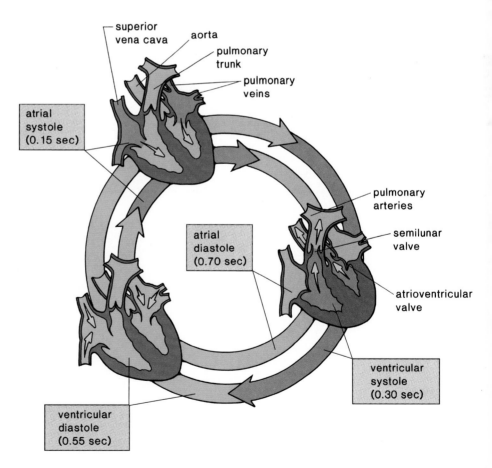

superior
vena cava

aorta

pulmonary
trunk

pulmonary
veins

atrial
systole
(0.15 sec)

atrial
diastole
(0.70 sec)

pulmonary
arteries

semilunar
valve

atrioventricular
valve

ventricular
systole
(0.30 sec)

ventricular
diastole
(0.55 sec)

Cardiac Cycle
Figure 5.5

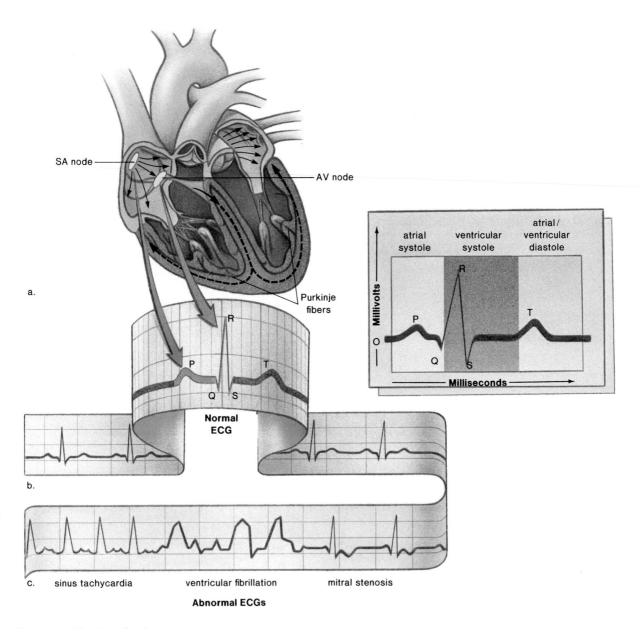

SA node

AV node

Purkinje
fibers

a.

atrial
systole

ventricular
systole

atrial/
ventricular
diastole

Millivolts

O

P

R

Q

S

T

Milliseconds

R

P

T

Q   S

Normal
ECG

b.

c.   sinus tachycardia   ventricular fibrillation   mitral stenosis

Abnormal ECGs

Control of Cardiac Cycle
Figure 5.6

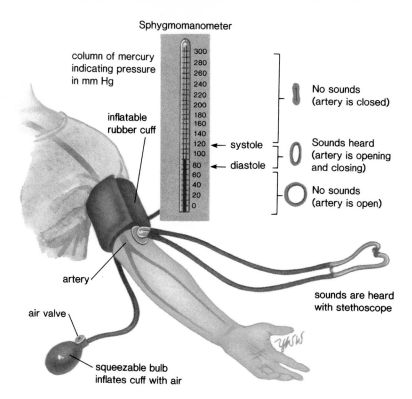

Sphygmomanometer

column of mercury
indicating pressure
in mm Hg

inflatable
rubber cuff

artery

air valve

squeezable bulb
inflates cuff with air

systole

diastole

No sounds
(artery is closed)

Sounds heard
(artery is opening
and closing)

No sounds
(artery is open)

sounds are heard
with stethoscope

Determination of Blood Pressure
Figure 5.7

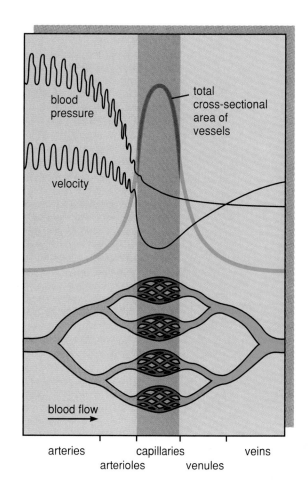

blood
pressure

velocity

total
cross-sectional
area of
vessels

blood flow

arteries
arterioles
capillaries
venules
veins

Blood Pressure and Velocity Diagram
Figure 5.8

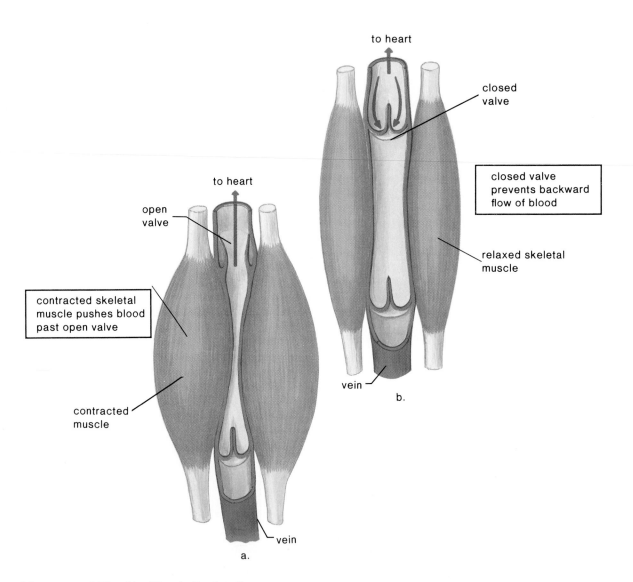

to heart

closed
valve

closed valve
prevents backward
flow of blood

relaxed skeletal
muscle

vein

b.

open
valve

to heart

contracted skeletal
muscle pushes blood
past open valve

contracted
muscle

vein

a.

Movement of Blood by Muscle Contraction
Figure 5.9

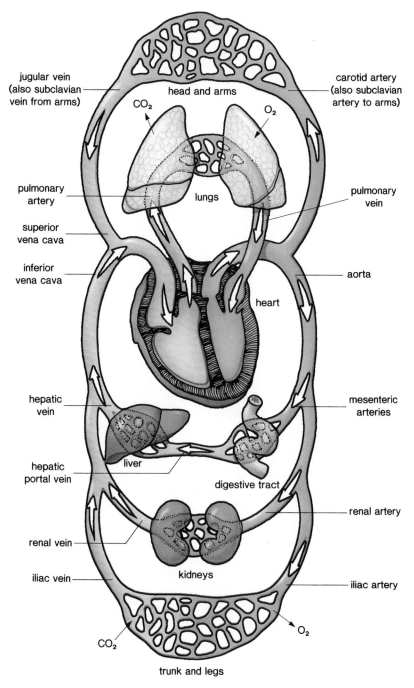

Cardiovascular System
Figure 5.10

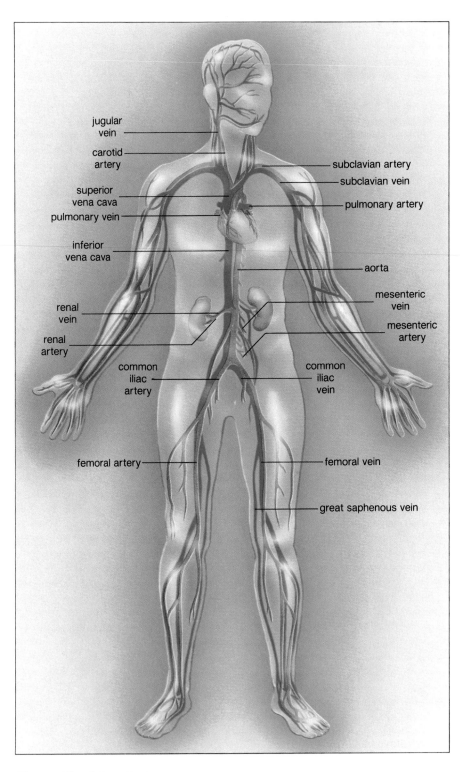

jugular
vein

carotid
artery

subclavian artery

subclavian vein

superior
vena cava

pulmonary artery

pulmonary vein

inferior
vena cava

aorta

mesenteric
vein

renal
vein

mesenteric
artery

renal
artery

common
iliac
artery

common
iliac
vein

femoral artery

femoral vein

great saphenous vein

Human Circulatory System
Figure 5.11

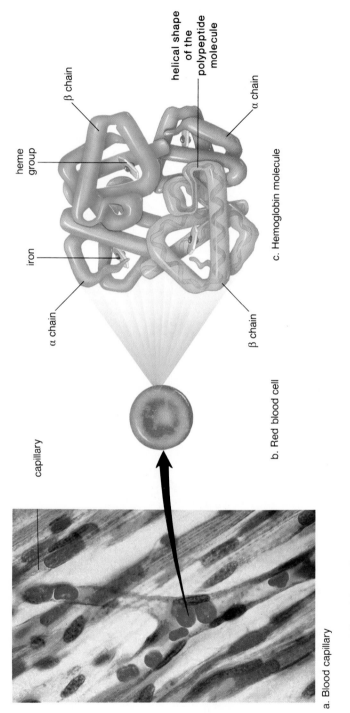

β chain

helical shape
of the
polypeptide
molecule

heme
group

α chain

iron

α chain

c. Hemoglobin molecule

β chain

capillary

b. Red blood cell

a. Blood capillary

Physiology of Red Blood Cells
Figure 6.2

41

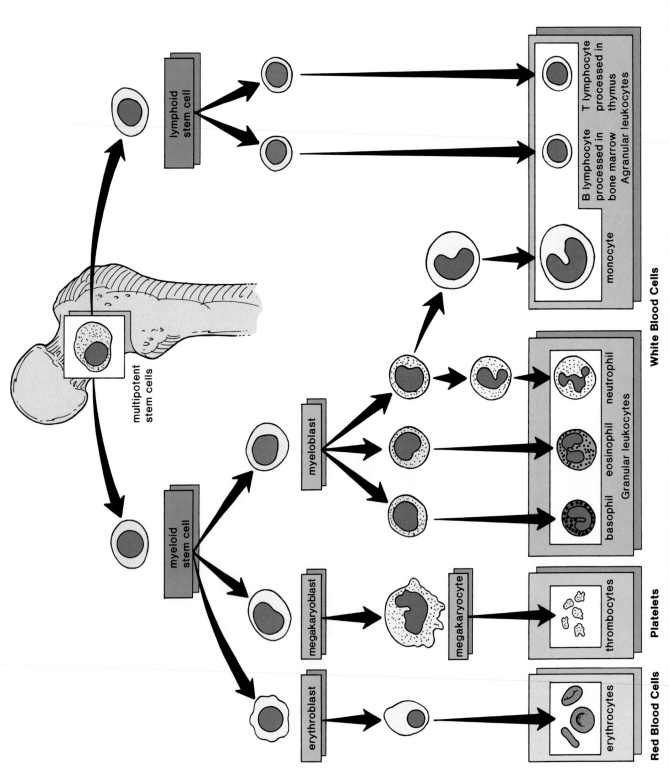

Blood Cell Formation
Figure 6.3

multipotent
stem cells

lymphoid
stem cell

myeloid
stem cell

myeloblast

megakaryoblast

megakaryocyte

erythroblast

T lymphocyte
processed in
thymus

B lymphocyte
processed in
bone marrow
Agranular leukocytes

monocyte

White Blood Cells

basophil   eosinophil   neutrophil
Granular leukocytes

thrombocytes

Platelets

erythrocytes

Red Blood Cells

42

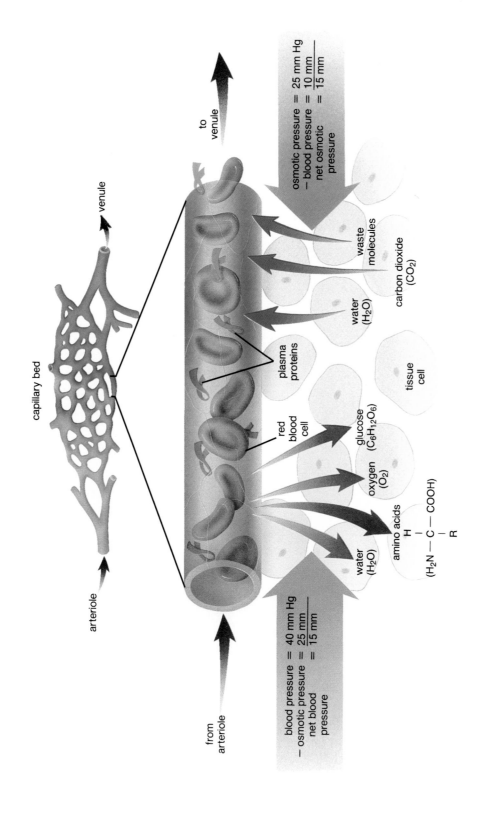

to
venule

venule

capillary bed

arteriole

osmotic pressure = 25 mm Hg
– blood pressure = 10 mm
net osmotic   = 15 mm
   pressure

waste
molecules

carbon dioxide
(CO₂)

water
(H₂O)

plasma
proteins

tissue
cell

red
blood
cell

glucose
(C₆H₁₂O₆)

oxygen
(O₂)

amino acids

(H₂N — C — COOH)
        |
        H
        |
        R

water
(H₂O)

from
arteriole

blood pressure   = 40 mm Hg
– osmotic pressure = 25 mm
net blood   = 15 mm
   pressure

Capillary Exchanges
Figure 6.7

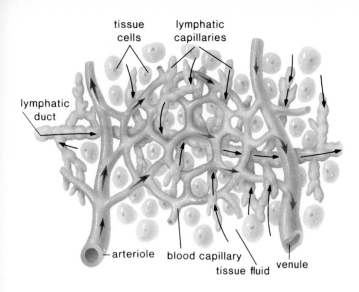

tissue cells

lymphatic capillaries

lymphatic duct

arteriole

blood capillary

tissue fluid

venule

Lymphatic Vessels
Figure 6.8

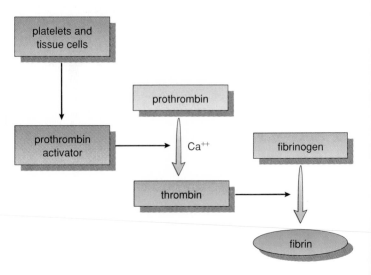

platelets and tissue cells

prothrombin

prothrombin activator

Ca^{++}

fibrinogen

thrombin

fibrin

Figure 6.9 a
Blood Clotting

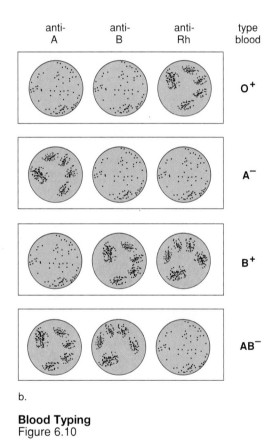

	anti- A	anti- B	anti- Rh	type blood

b.

Blood Typing
Figure 6.10

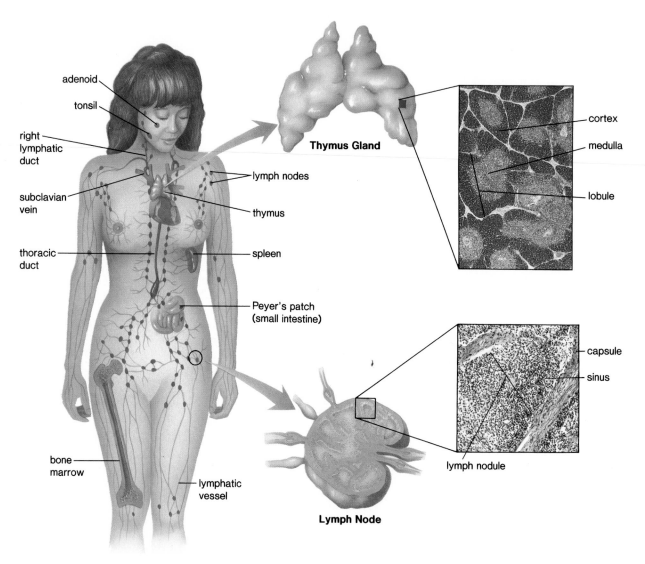

adenoid
tonsil
right lymphatic duct
subclavian vein
thoracic duct
bone marrow
lymphatic vessel

lymph nodes
thymus
spleen
Peyer's patch (small intestine)

Thymus Gland

cortex
medulla
lobule

capsule
sinus
lymph nodule

Lymph Node

Lymphatic System
Figure 7.1

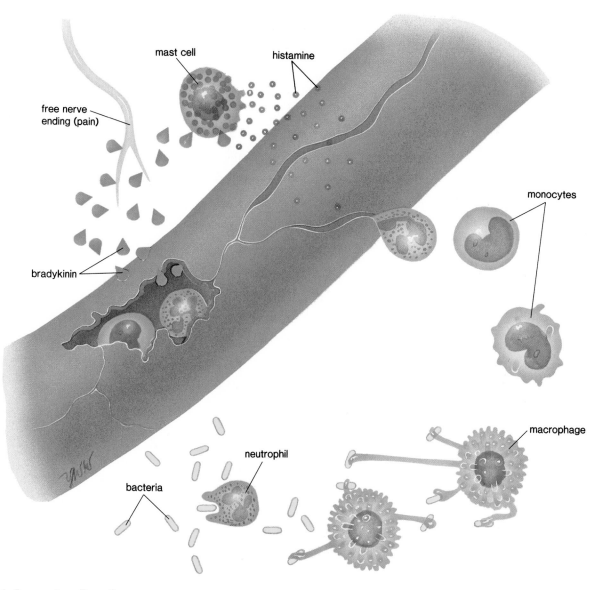

Inflammatory Reaction
Figure 7.3

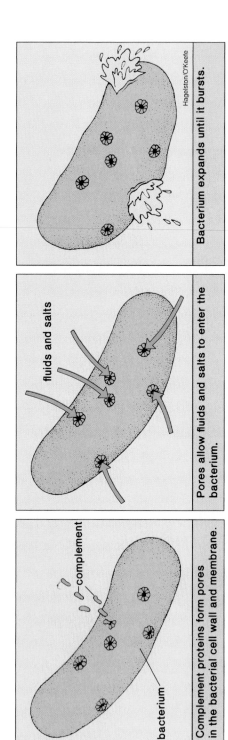

Complement System Actions
Figure 7.4

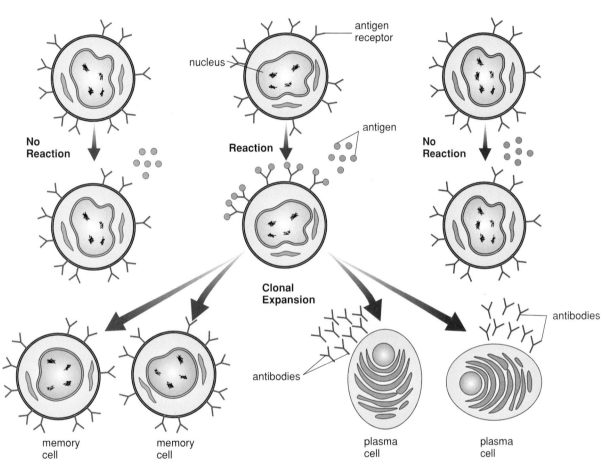

antigen receptor

nucleus

antigen

No Reaction

Reaction

No Reaction

Clonal Expansion

antibodies

antibodies

memory cell

memory cell

plasma cell

plasma cell

Clonal Selection Theory
Figure 7.5

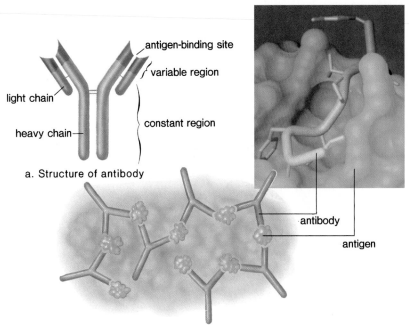

a. Structure of antibody

antigen-binding site

variable region

light chain

heavy chain

constant region

antibody

antigen

b. Antigen-antibody complex

Antigen-Antibody Reaction
Figure 7.6

Cell-Mediated Immunity
Figure 7.7

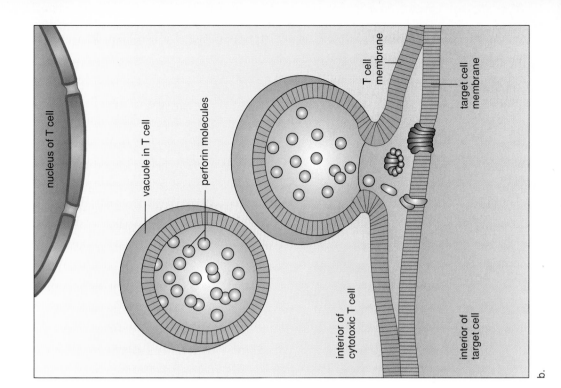

nucleus of T cell

vacuole in T cell

perforin molecules

T cell
membrane

target cell
membrane

interior of
cytotoxic T cell

interior of
target cell

b.

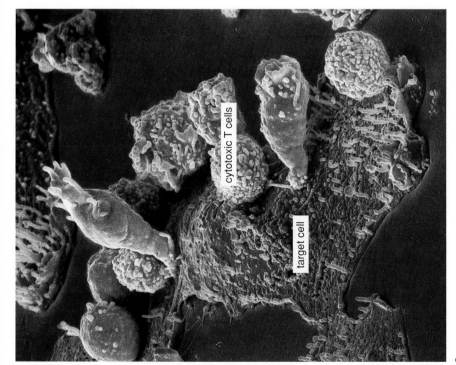

cytotoxic T cells

target cell

a.

51

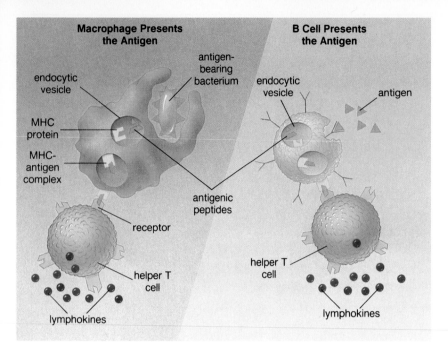

Macrophage Presents the Antigen

endocytic vesicle

antigen-bearing bacterium

MHC protein

MHC-antigen complex

antigenic peptides

receptor

helper T cell

lymphokines

B Cell Presents the Antigen

endocytic vesicle

antigen

helper T cell

lymphokines

a.

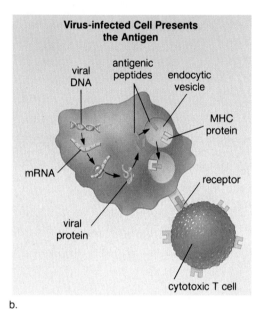

Virus-infected Cell Presents the Antigen

viral DNA

antigenic peptides

endocytic vesicle

MHC protein

mRNA

receptor

viral protein

cytotoxic T cell

b.

T Cell Activation
Figure 7.8

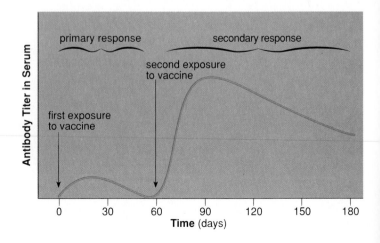

primary response

secondary response

second exposure to vaccine

first exposure to vaccine

Antibody Titer in Serum

Time (days)

Development of Active Immunity
Figure 7.9

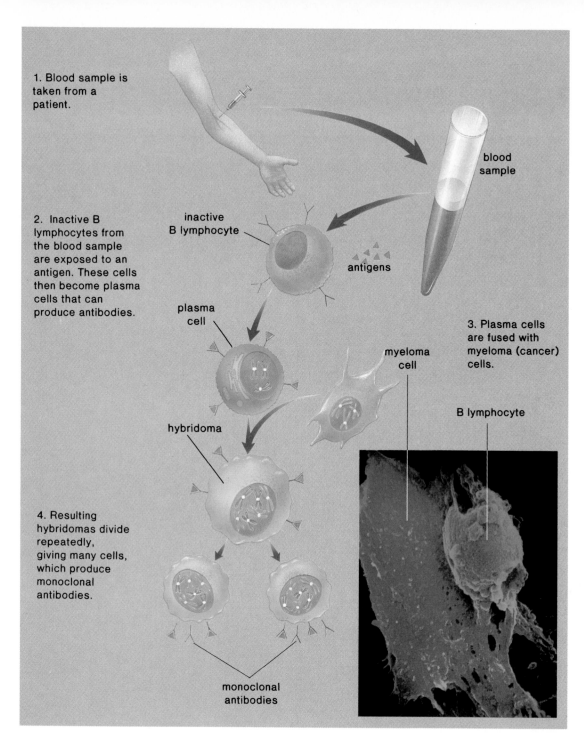

1. Blood sample is taken from a patient.

blood sample

2. Inactive B lymphocytes from the blood sample are exposed to an antigen. These cells then become plasma cells that can produce antibodies.

inactive B lymphocyte

antigens

plasma cell

3. Plasma cells are fused with myeloma (cancer) cells.

myeloma cell

B lymphocyte

hybridoma

4. Resulting hybridomas divide repeatedly, giving many cells, which produce monoclonal antibodies.

monoclonal antibodies

Monoclonal Antibody Reaction
Figure 7.10

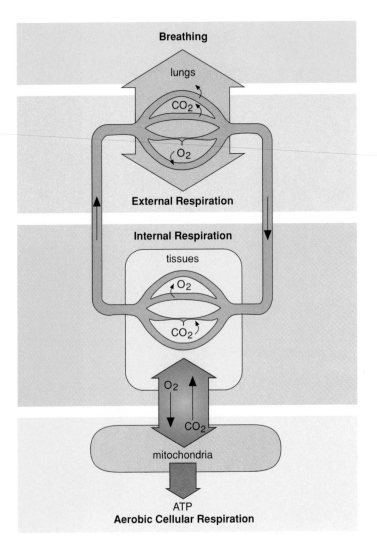

Respiration
Figure 8.1

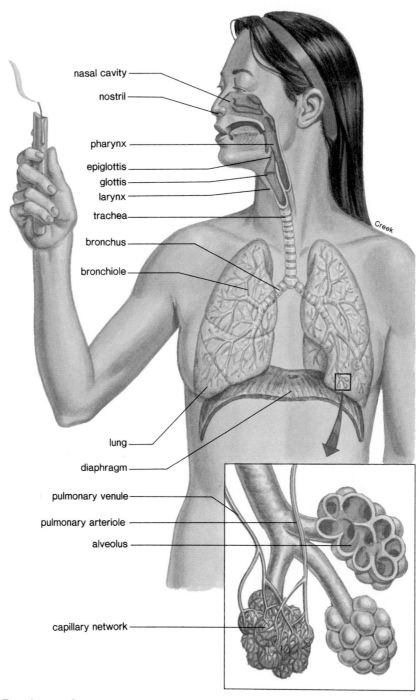

nasal cavity

nostril

pharynx

epiglottis

glottis

larynx

trachea

bronchus

bronchiole

Creek

lung

diaphragm

pulmonary venule

pulmonary arteriole

alveolus

capillary network

Respiratory System
Figure 8.2

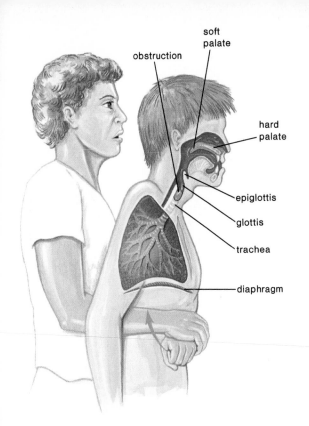

soft
palate

obstruction

hard
palate

epiglottis

glottis

trachea

diaphragm

Heimlich Maneuver
Figure 8.3

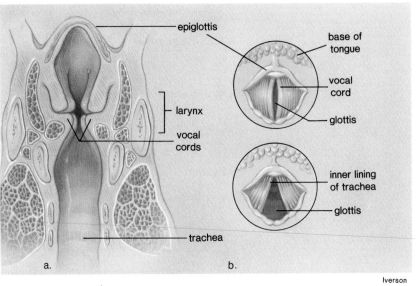

epiglottis

base of
tongue

vocal
cord

larynx

glottis

vocal
cords

inner lining
of trachea

glottis

trachea

a.

b.

Iverson

Vocal Cords
Figure 8.4

These structures are to be associated with inspiration.

These structures are to be associated with expiration.

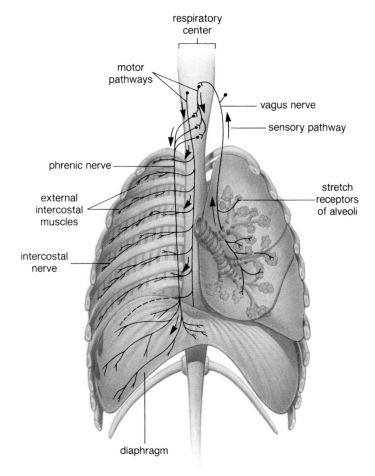

respiratory center

motor pathways

vagus nerve

sensory pathway

phrenic nerve

external intercostal muscles

stretch receptors of alveoli

intercostal nerve

diaphragm

Nervous Control of Breathing
Figure 8.6

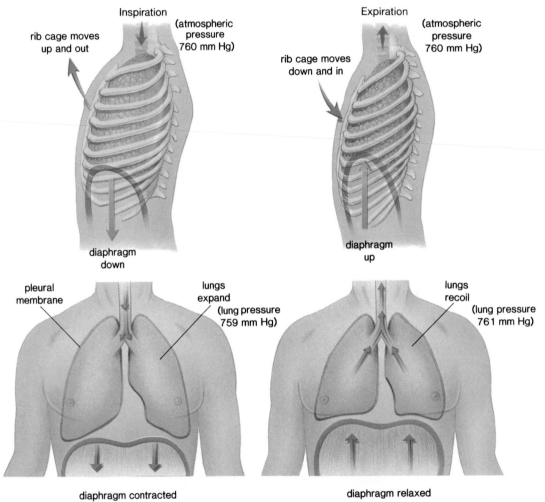

Inspiration

rib cage moves
up and out

(atmospheric
pressure
760 mm Hg)

diaphragm
down

pleural
membrane

lungs
expand
(lung pressure
759 mm Hg)

diaphragm contracted

Pressure in lungs decreases:
air comes rushing in

Expiration

rib cage moves
down and in

(atmospheric
pressure
760 mm Hg)

diaphragm
up

lungs
recoil
(lung pressure
761 mm Hg)

diaphragm relaxed

Pressure in lungs increases:
air is pushed out

Inspiration and Expiration
Figure 8.7

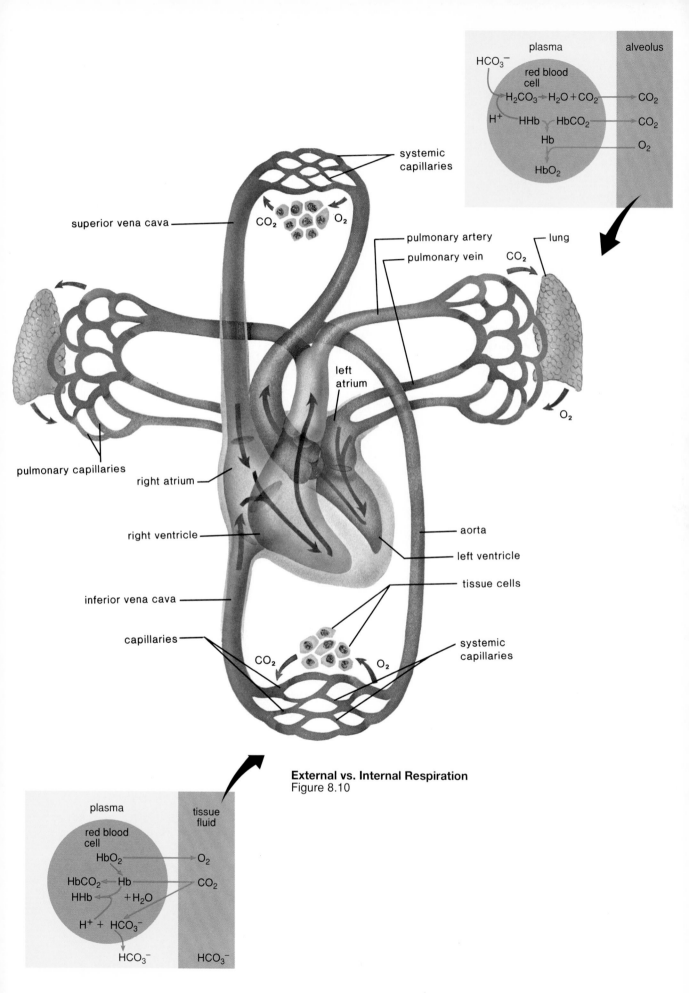

External vs. Internal Respiration
Figure 8.10

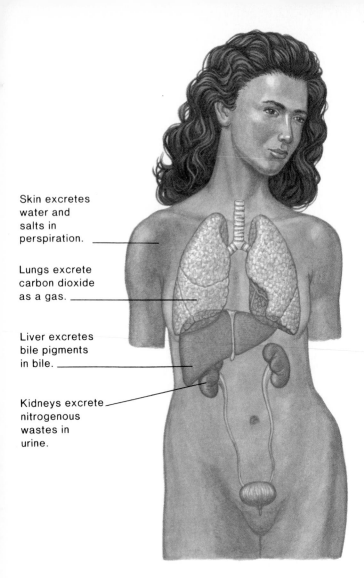

Skin excretes
water and
salts in
perspiration.

Lungs excrete
carbon dioxide
as a gas.

Liver excretes
bile pigments
in bile.

Kidneys excrete
nitrogenous
wastes in
urine.

Organs of Excretion
Figure 9.1

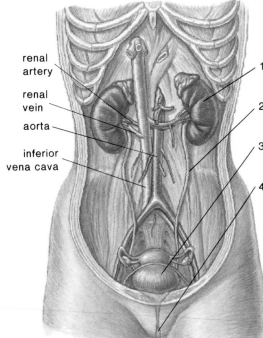

renal
artery

renal
vein

aorta

inferior
vena cava

1. Kidneys produce
   urine.

2. Ureters transport
   urine.

3. Urinary bladder
   stores urine.

4. Urethra passes
   urine to outside.

Urinary System
Figure 9.2

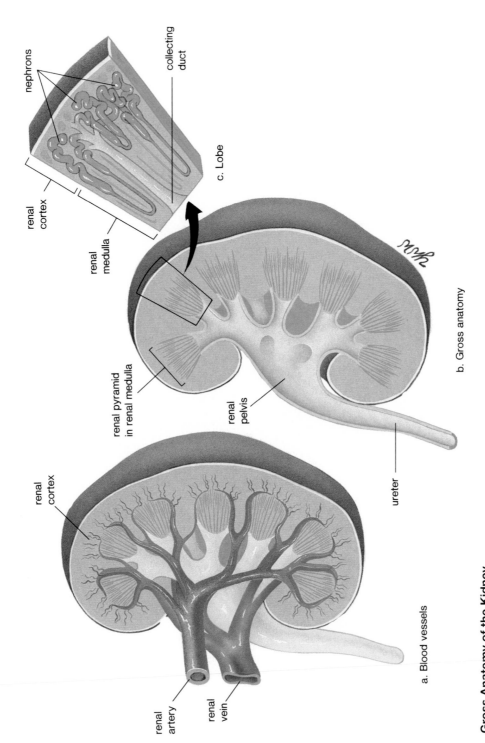

nephrons

collecting
duct

c. Lobe

renal
cortex

renal
medulla

renal pyramid
in renal medulla

renal
pelvis

b. Gross anatomy

ureter

renal
cortex

renal
artery

renal
vein

a. Blood vessels

Gross Anatomy of the Kidney
Figure 9.3

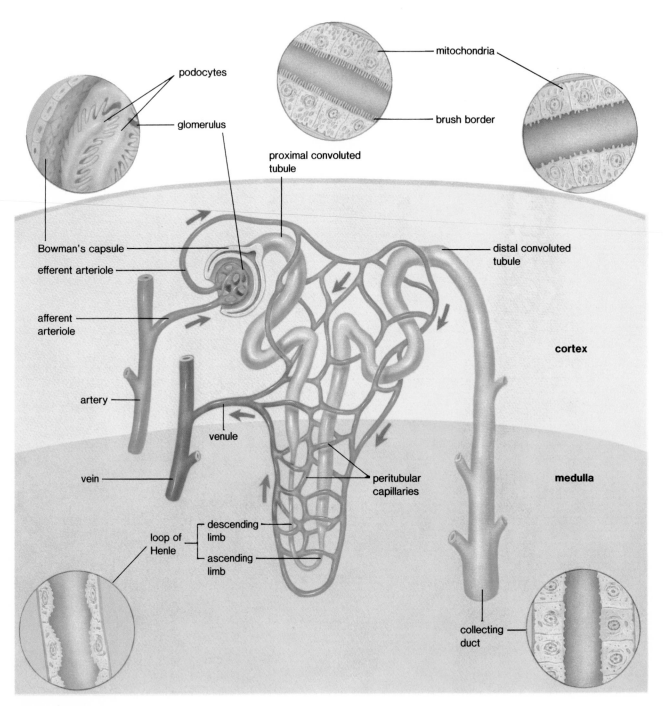

podocytes

glomerulus

mitochondria

brush border

proximal convoluted
tubule

Bowman's capsule

efferent arteriole

distal convoluted
tubule

afferent
arteriole

cortex

artery

venule

vein

peritubular
capillaries

medulla

loop of
Henle

descending
limb

ascending
limb

collecting
duct

Nephron Anatomy
Figure 9.4

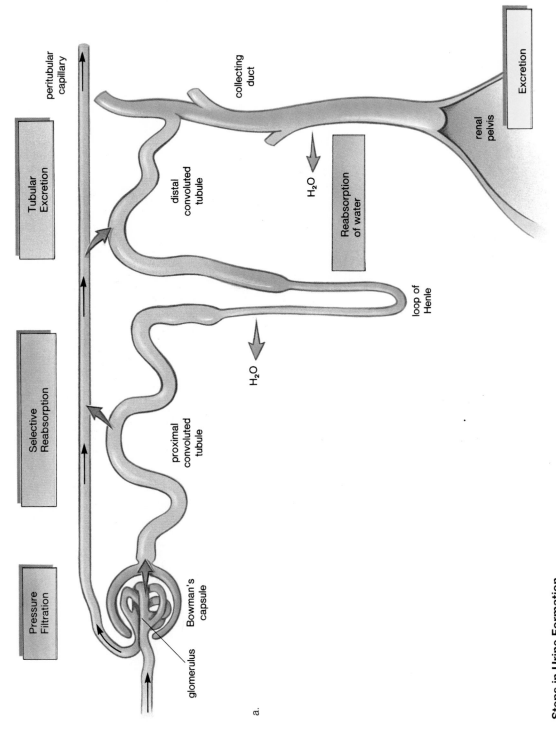

peritubular capillary

Tubular Excretion

Selective Reabsorption

Pressure Filtration

glomerulus

Bowman's capsule

proximal convoluted tubule

H_2O

loop of Henle

distal convoluted tubule

collecting duct

H_2O

Reabsorption of water

renal pelvis

Excretion

a.

Steps in Urine Formation
Figure 9.5 a

63

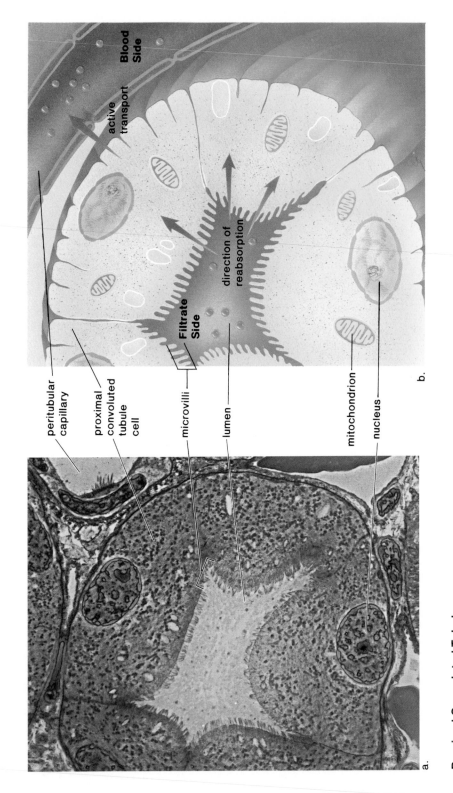

Blood Side

active transport

direction of reabsorption

Filtrate Side

b.

peritubular capillary

proximal convoluted tubule cell

microvilli

lumen

mitochondrion

nucleus

a.

Proximal Convoluted Tubule
Figure 9.6

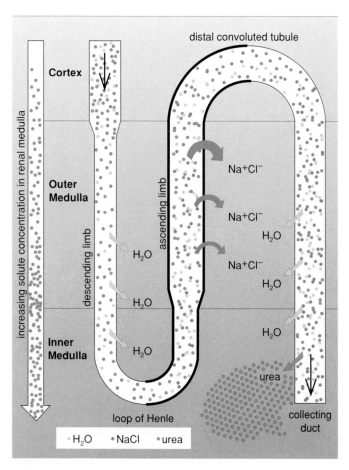

Loop of Henle
Figure 9.7

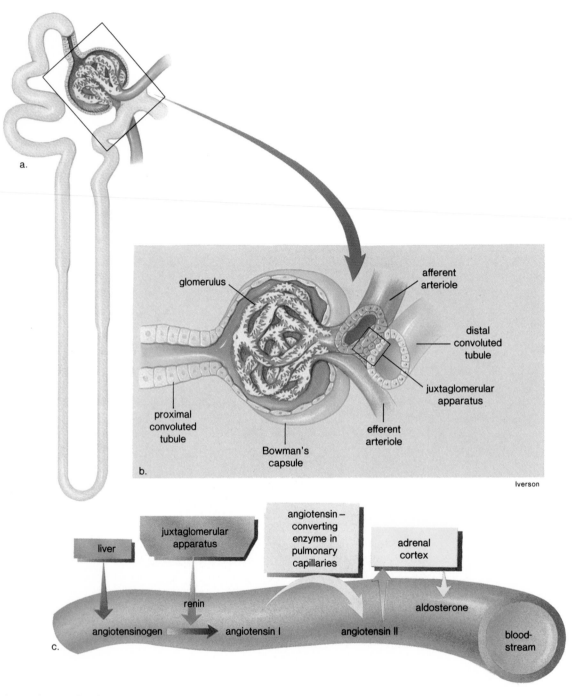

a.

glomerulus

afferent
arteriole

distal
convoluted
tubule

juxtaglomerular
apparatus

proximal
convoluted
tubule

efferent
arteriole

Bowman's
capsule

b.

Iverson

liver

juxtaglomerular
apparatus

angiotensin –
converting
enzyme in
pulmonary
capillaries

adrenal
cortex

renin

aldosterone

angiotensinogen

angiotensin I

angiotensin II

blood-
stream

c.

Juxtaglomerular Apparatus
Figure 9.8

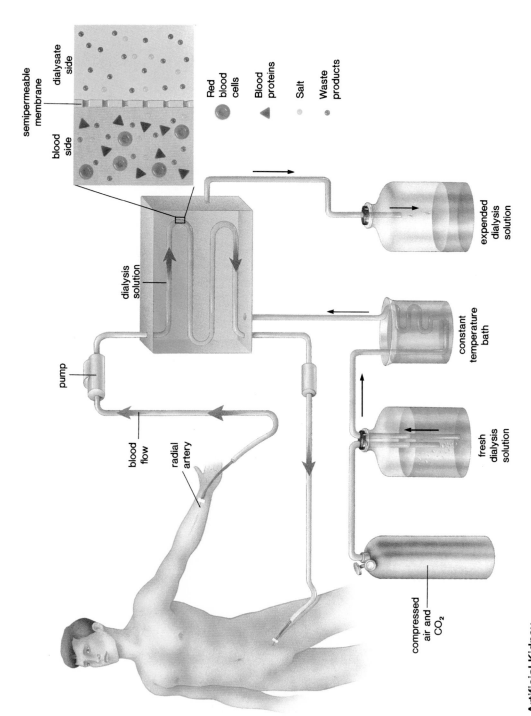

Artificial Kidney
Figure 9.9

semipermeable membrane

dialysate side

blood side

Red blood cells

Blood proteins

Salt

Waste products

dialysis solution

expended dialysis solution

constant temperature bath

fresh dialysis solution

compressed air and CO_2

pump

blood flow

radial artery

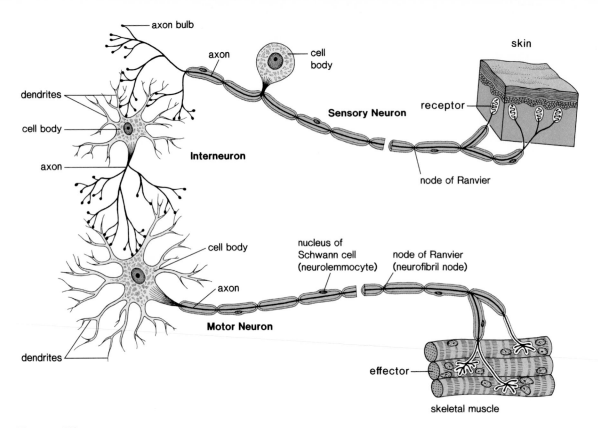

Types of Neurons
Figure 10.1

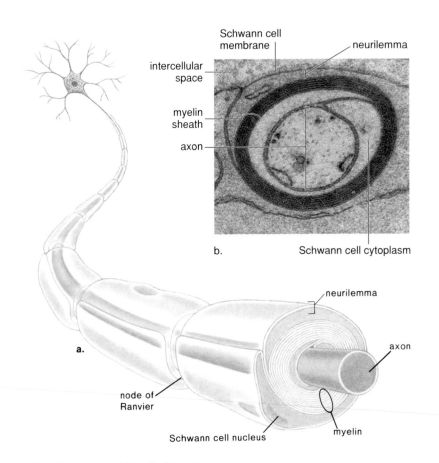

Neurilemma and Myelin Sheath
Figure 10.2

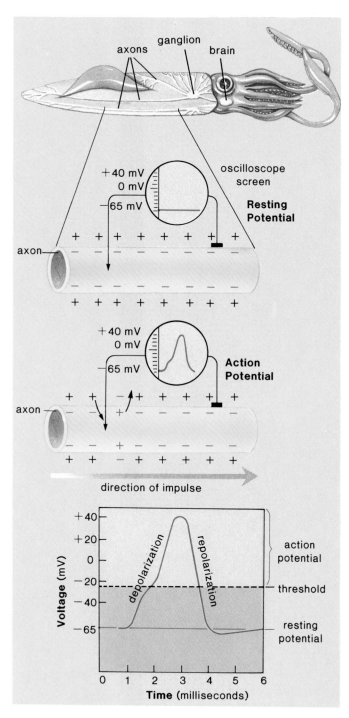

Nerve Impulse
Figure 10.3

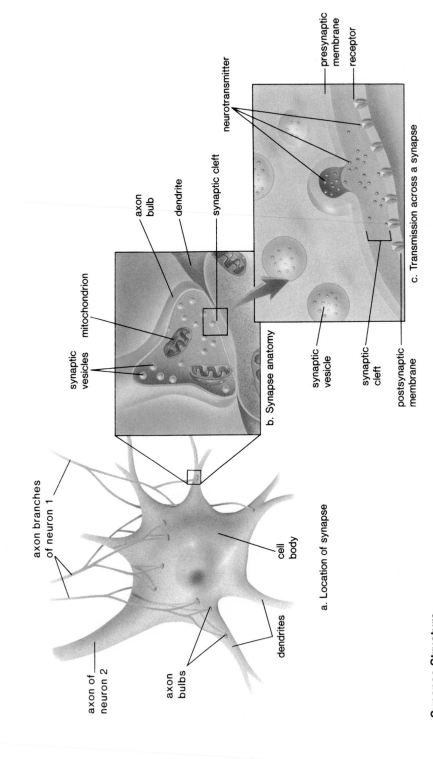

axon branches
of neuron 1

axon of
neuron 2

axon
bulbs

dendrites

cell
body

a. Location of synapse

mitochondrion

synaptic
vesicles

axon
bulb

dendrite

synaptic cleft

b. Synapse anatomy

synaptic
vesicle

synaptic
cleft

postsynaptic
membrane

neurotransmitter

presynaptic
membrane

receptor

c. Transmission across a synapse

Synapse Structure
Figure 10.5

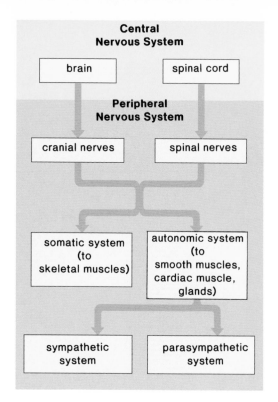

Organization of Nervous System
Figure 10.6 b

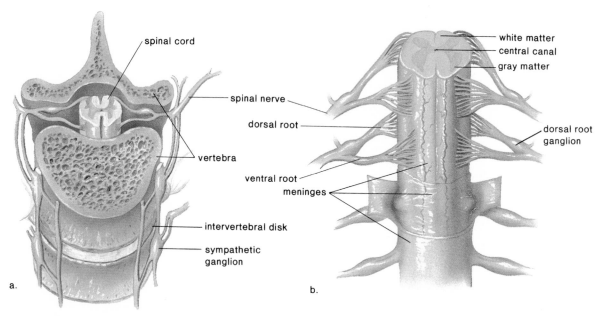

a.

b.

Spinal Cord Anatomy
Figure 10.7

71

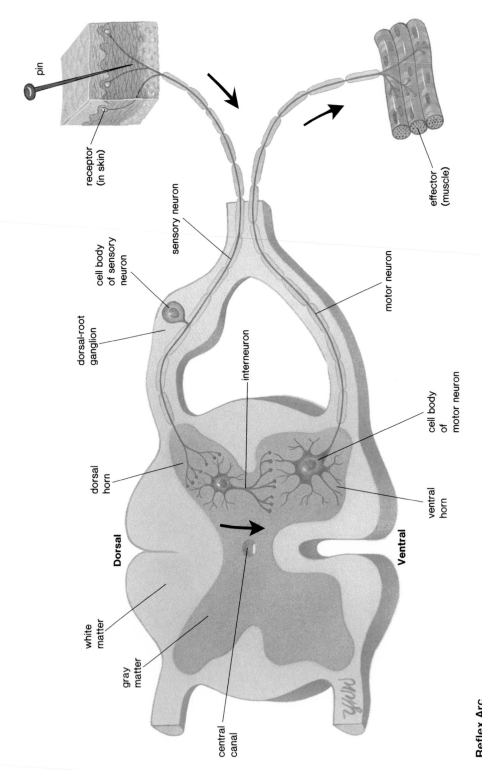

pin

receptor
(in skin)

sensory neuron

cell body
of sensory
neuron

dorsal-root
ganglion

interneuron

Dorsal

dorsal
horn

white
matter

gray
matter

central
canal

effector
(muscle)

motor neuron

cell body
of
motor neuron

ventral
horn

Ventral

Reflex Arc
Figure 10.8

72

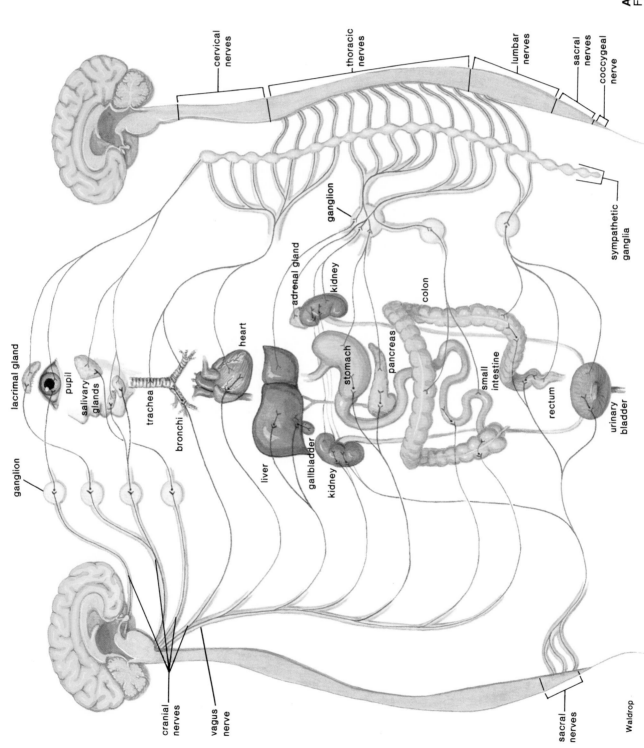

Autonomic Nervous System
Figure 10.9

cervical nerves

thoracic nerves

lumbar nerves

sacral nerves

coccygeal nerve

ganglion

sympathetic ganglia

adrenal gland

kidney

colon

stomach

pancreas

small intestine

rectum

urinary bladder

heart

liver

gallbladder

kidney

trachea

bronchi

lacrimal gland

pupil

salivary glands

ganglion

cranial nerves

vagus nerve

sacral nerves

Waldrop.

73

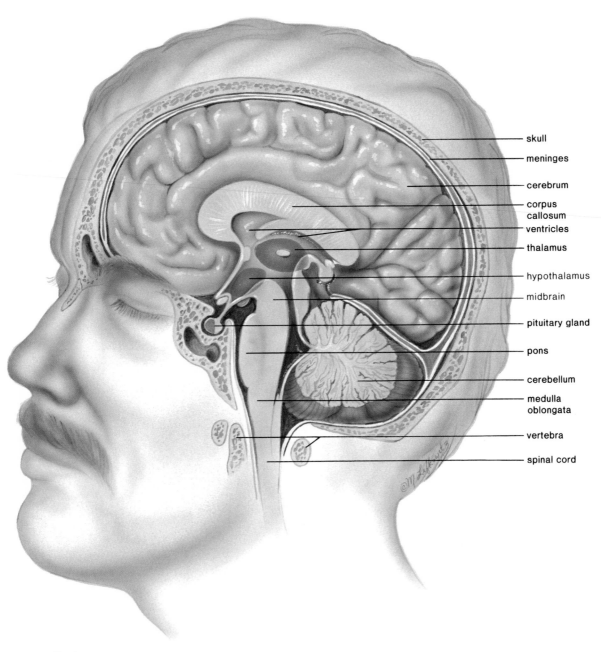

skull

meninges

cerebrum

corpus
callosum

ventricles

thalamus

hypothalamus

midbrain

pituitary gland

pons

cerebellum

medulla
oblongata

vertebra

spinal cord

Human Brain
Figure 10.10

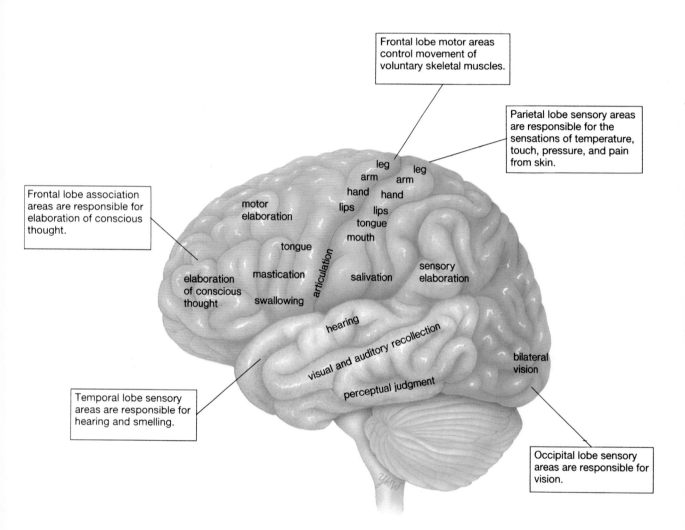

Frontal lobe motor areas control movement of voluntary skeletal muscles.

Parietal lobe sensory areas are responsible for the sensations of temperature, touch, pressure, and pain from skin.

Frontal lobe association areas are responsible for elaboration of conscious thought.

leg
leg
arm
arm
hand
hand
motor
elaboration
lips
lips
tongue
mouth
tongue
articulation
mastication
salivation
sensory
elaboration
elaboration
of conscious
thought
swallowing
hearing
visual and auditory recollection
bilateral
vision
perceptual judgment

Temporal lobe sensory areas are responsible for hearing and smelling.

Occipital lobe sensory areas are responsible for vision.

Lobes of the Brain
Figure 10.12

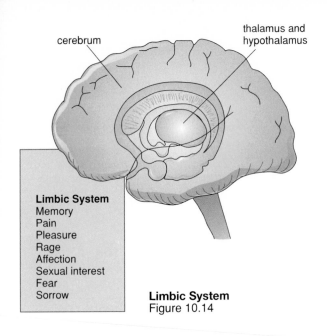

cerebrum

thalamus and
hypothalamus

Limbic System
Memory
Pain
Pleasure
Rage
Affection
Sexual interest
Fear
Sorrow

Limbic System
Figure 10.14

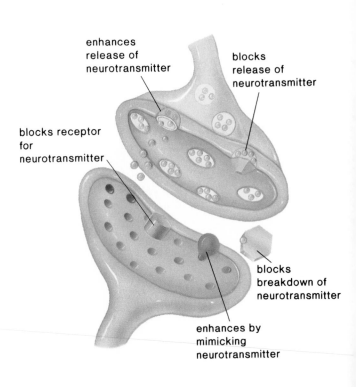

enhances
release of
neurotransmitter

blocks
release of
neurotransmitter

blocks receptor
for
neurotransmitter

blocks
breakdown of
neurotransmitter

enhances by
mimicking
neurotransmitter

Drug Action at Synapses
Figure 10.15

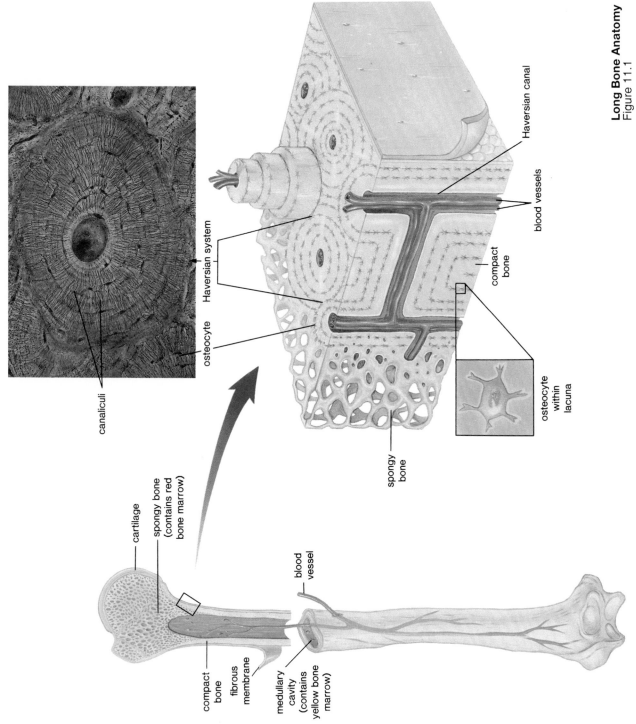

canaliculi

osteocyte

Haversian system

Haversian canal

blood vessels

compact bone

osteocyte within lacuna

spongy bone

cartilage

spongy bone (contains red bone marrow)

compact bone

fibrous membrane

blood vessel

medullary cavity (contains yellow bone marrow)

Long Bone Anatomy
Figure 11.1

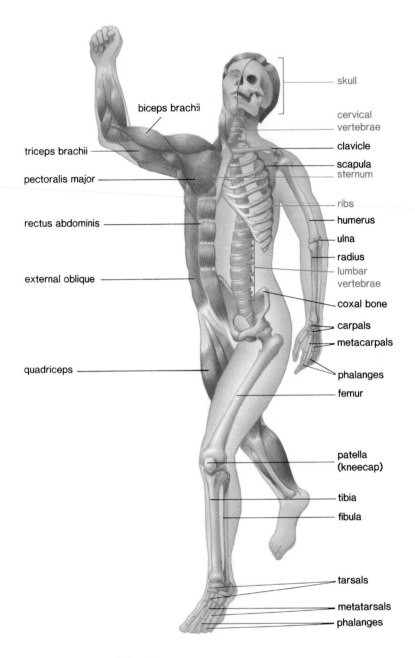

skull

cervical
vertebrae

biceps brachii

clavicle

triceps brachii

scapula

pectoralis major

sternum

ribs

rectus abdominis

humerus

ulna

radius

lumbar
vertebrae

external oblique

coxal bone

carpals

metacarpals

phalanges

quadriceps

femur

patella
(kneecap)

tibia

fibula

tarsals

metatarsals

phalanges

Major Bones and Muscles
Figure 11.2

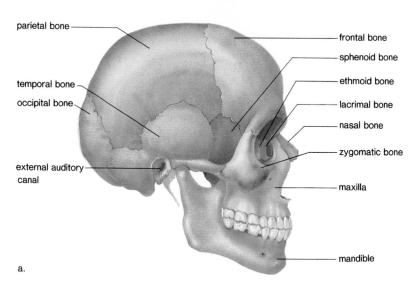

parietal bone

frontal bone

sphenoid bone

ethmoid bone

lacrimal bone

temporal bone

occipital bone

nasal bone

zygomatic bone

external auditory
canal

maxilla

mandible

a.

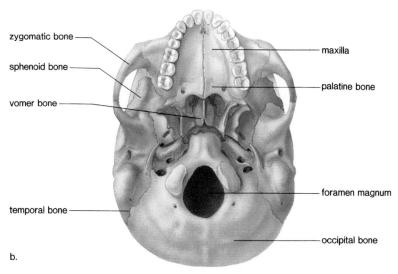

zygomatic bone

maxilla

sphenoid bone

palatine bone

vomer bone

temporal bone

foramen magnum

occipital bone

b.

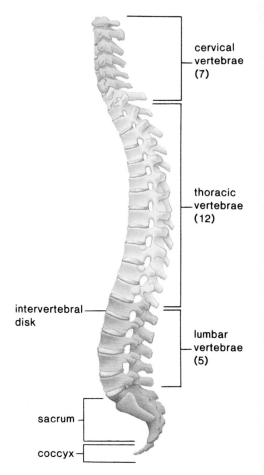

cervical
vertebrae
(7)

thoracic
vertebrae
(12)

intervertebral
disk

lumbar
vertebrae
(5)

sacrum

coccyx

The Vertebral Column
Figure 11.4

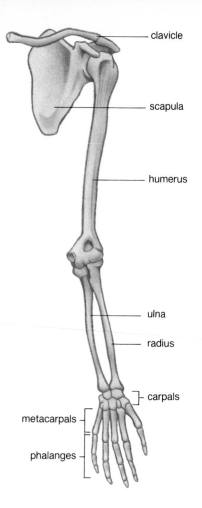

clavicle

scapula

humerus

ulna

radius

carpals

metacarpals

phalanges

Bones of the Pectoral Girdle
Figure 11.5

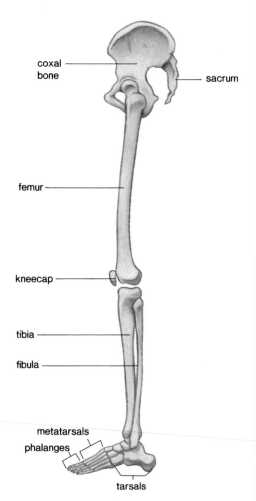

coxal
bone

sacrum

femur

kneecap

tibia

fibula

metatarsals

phalanges

tarsals

Bones of the Pelvic Girdle
Figure 11.6

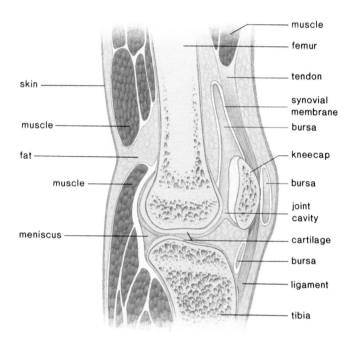

muscle

femur

tendon

synovial
membrane

skin

bursa

muscle

kneecap

fat

bursa

muscle

joint
cavity

cartilage

meniscus

bursa

ligament

tibia

Knee Joint
Figure 11.7

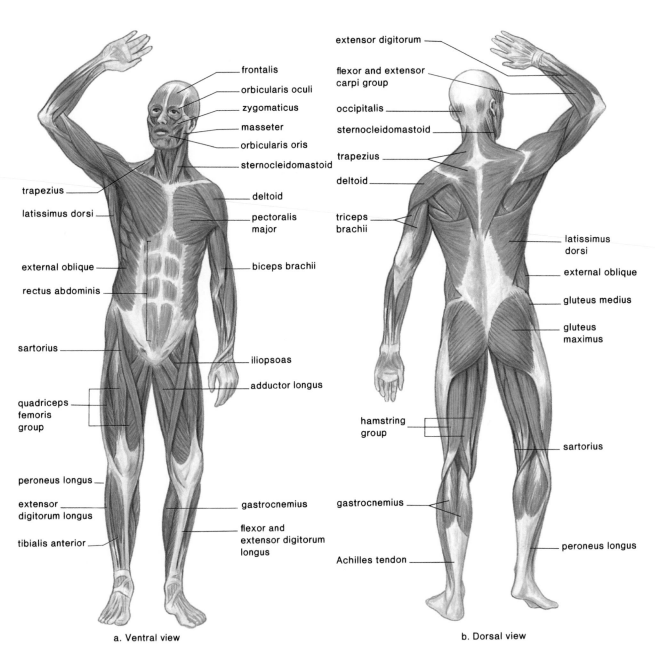

frontalis
orbicularis oculi
zygomaticus
masseter
orbicularis oris
sternocleidomastoid

trapezius
latissimus dorsi

external oblique
rectus abdominis

sartorius

quadriceps
femoris
group

peroneus longus
extensor
digitorum longus
tibialis anterior

deltoid
pectoralis
major

biceps brachii

iliopsoas
adductor longus

gastrocnemius
flexor and
extensor digitorum
longus

a. Ventral view

extensor digitorum
flexor and extensor
carpi group
occipitalis
sternocleidomastoid
trapezius
deltoid
triceps
brachii

latissimus
dorsi
external oblique
gluteus medius
gluteus
maximus

hamstring
group

gastrocnemius

Achilles tendon

sartorius

peroneus longus

b. Dorsal view

Superficial Skeletal Muscles, Ventral View
Figure 11.9 a

Superficial Skeletal Muscles, Dorsal View
Figure 11.9 b

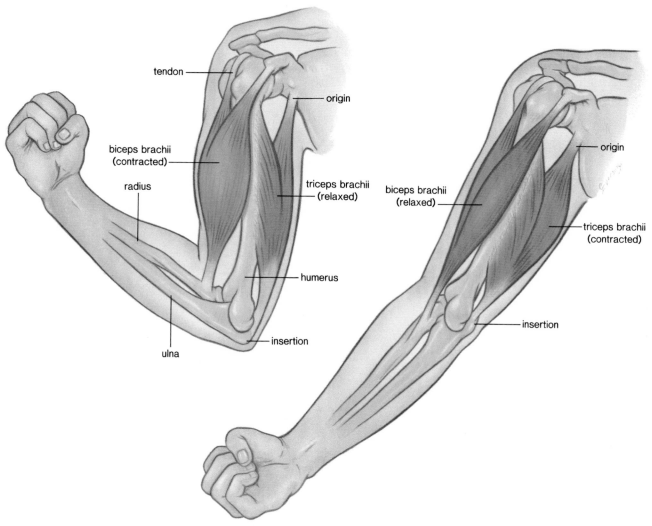

tendon

origin

biceps brachii
(contracted)

radius

triceps brachii
(relaxed)

biceps brachii
(relaxed)

origin

triceps brachii
(contracted)

humerus

insertion

insertion

ulna

Biceps and Triceps
Figure 11.10

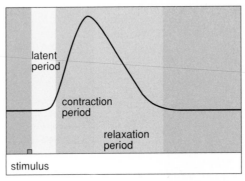

b.

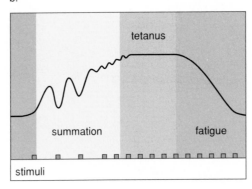

c.

Physiology of Skeletal Muscle Contraction
Figure 11.11 b,c

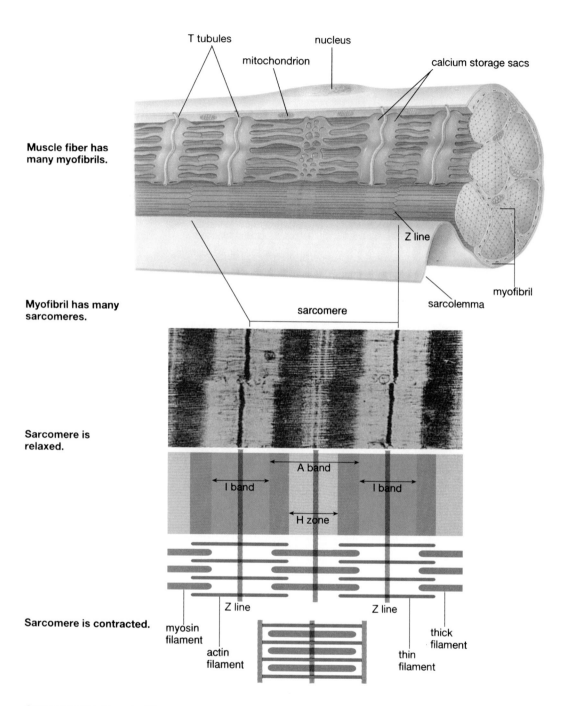

T tubules   nucleus

mitochondrion

calcium storage sacs

Muscle fiber has many myofibrils.

Z line

myofibril

sarcolemma

Myofibril has many sarcomeres.

sarcomere

Sarcomere is relaxed.

A band

I band   I band

H zone

Z line   Z line

Sarcomere is contracted.

myosin filament

actin filament

thick filament

thin filament

Anatomy of a Muscle Fiber
Figure 11.12

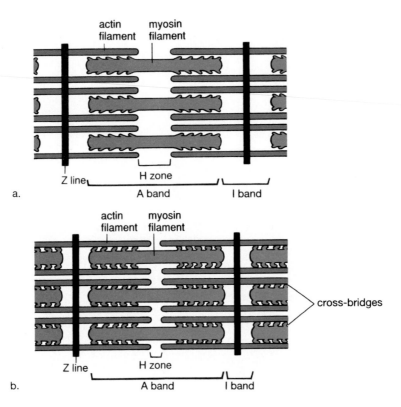

Sliding Filament Theory
Figure 11.13

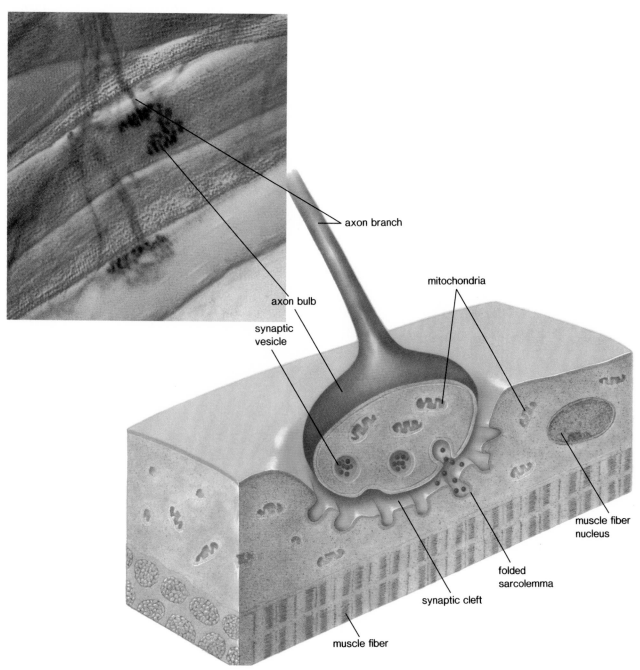

axon branch

mitochondria

axon bulb

synaptic
vesicle

muscle fiber
nucleus

folded
sarcolemma

synaptic cleft

muscle fiber

Muscle Innervation
Figure 11.14

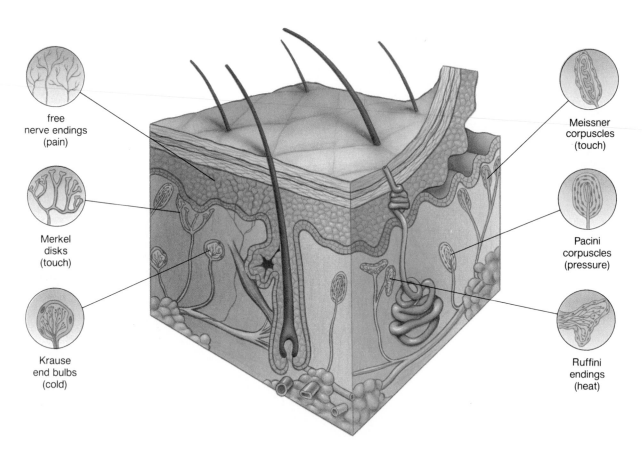

free
nerve endings
(pain)

Merkel
disks
(touch)

Krause
end bulbs
(cold)

Meissner
corpuscles
(touch)

Pacini
corpuscles
(pressure)

Ruffini
endings
(heat)

Receptors in Human Skin
Figure 12.1

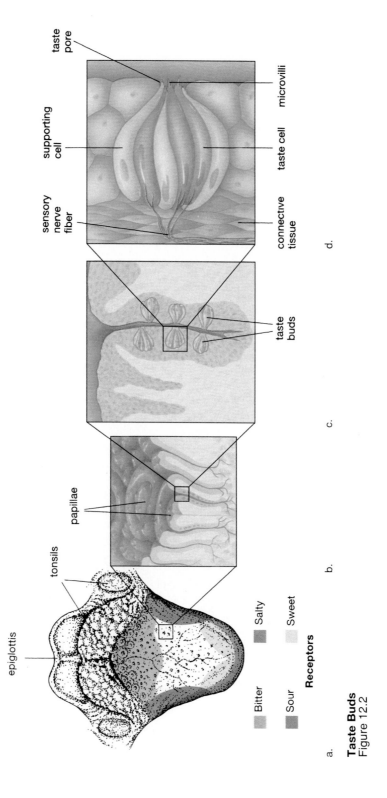

taste
pore

microvilli

supporting
cell

taste cell

sensory
nerve
fiber

connective
tissue

d.

taste
buds

c.

papillae

b.

epiglottis

tonsils

Bitter

Salty

Sour

Sweet

Receptors

a.

Taste Buds
Figure 12.2

89

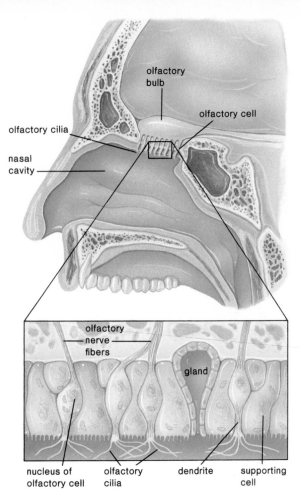

Olfactory Cell Anatomy
Figure 12.3

olfactory
bulb

olfactory cell

olfactory cilia

nasal
cavity

olfactory
nerve
fibers

gland

nucleus of
olfactory cell

olfactory
cilia

dendrite

supporting
cell

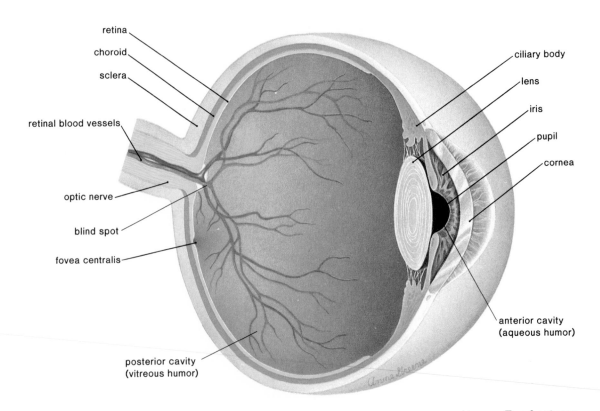

retina

choroid

sclera

retinal blood vessels

optic nerve

blind spot

fovea centralis

posterior cavity
(vitreous humor)

ciliary body

lens

iris

pupil

cornea

anterior cavity
(aqueous humor)

Human Eye Anatomy
Figure 12.4

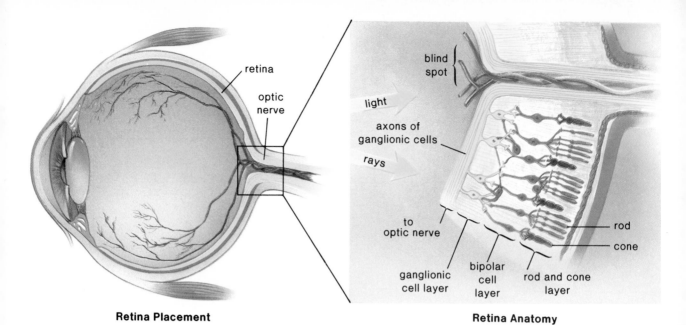

Retina Placement

Retina Anatomy

Retina Anatomy
Figure 12.5

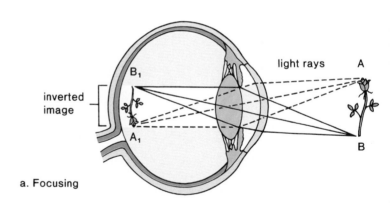

a. Focusing

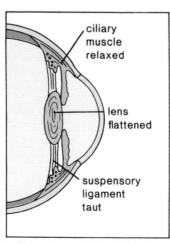

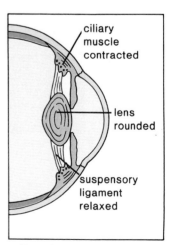

Focusing
Figure 12.6

b. Focusing on distant object

c. Focusing on near object

Structure of Rods and Cones
Figure 12.8

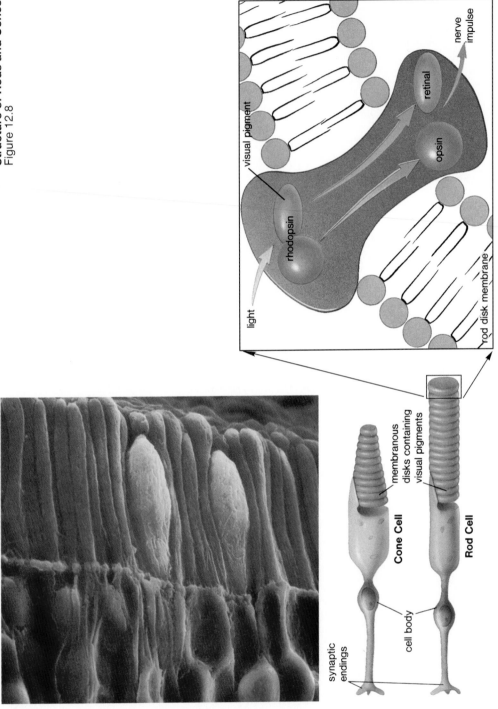

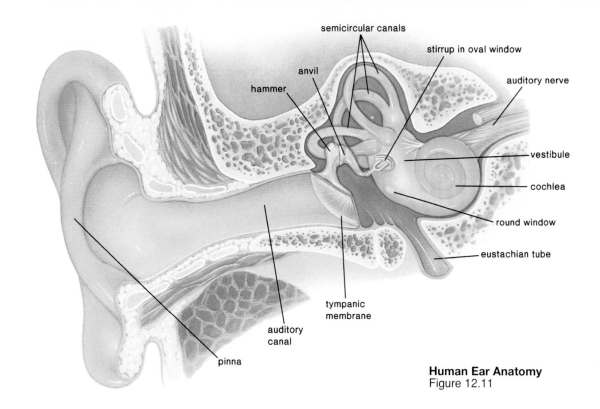

Human Ear Anatomy
Figure 12.11

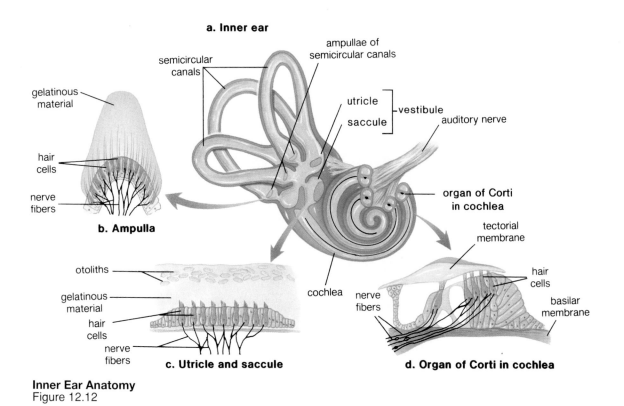

a. Inner ear

gelatinous material

hair cells

nerve fibers

b. Ampulla

semicircular canals

ampullae of semicircular canals

utricle

saccule

vestibule

auditory nerve

organ of Corti in cochlea

tectorial membrane

hair cells

basilar membrane

cochlea

nerve fibers

otoliths

gelatinous material

hair cells

nerve fibers

c. Utricle and saccule

d. Organ of Corti in cochlea

Inner Ear Anatomy
Figure 12.12

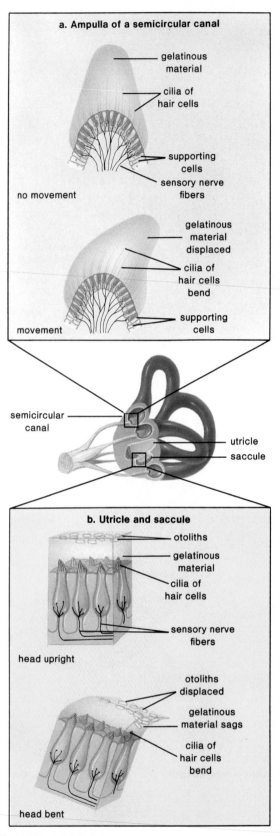

Receptors for Balance
Figure 12.13

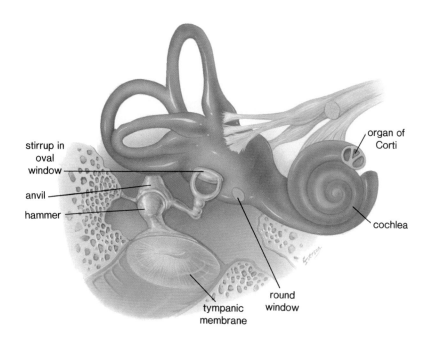

stirrup in oval window

anvil

hammer

organ of Corti

cochlea

tympanic membrane

round window

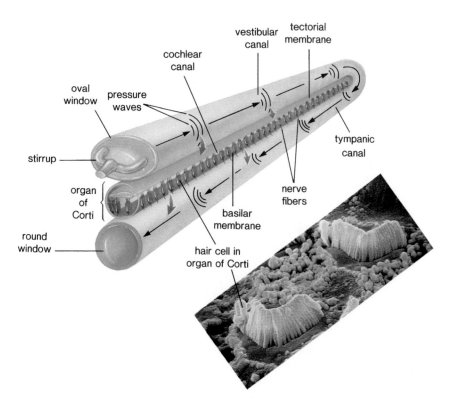

oval window

stirrup

organ of Corti

round window

pressure waves

cochlear canal

vestibular canal

tectorial membrane

tympanic canal

nerve fibers

basilar membrane

hair cell in organ of Corti

Organ of Corti
Figure 12.14

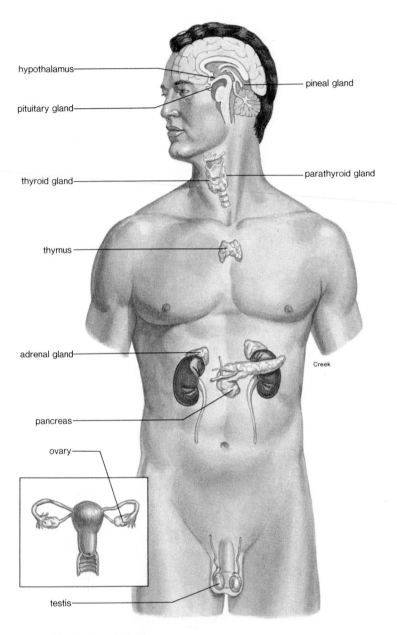

hypothalamus

pineal gland

pituitary gland

thyroid gland

parathyroid gland

thymus

adrenal gland

Creek

pancreas

ovary

testis

Major Endocrine Glands
Figure 13.1

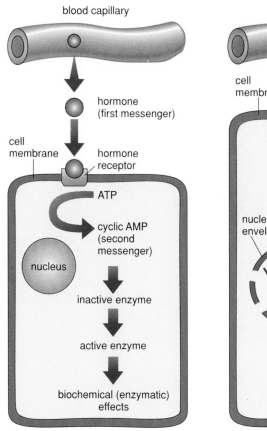

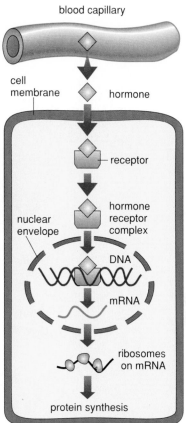

a.   b.

Cellular Activity of Hormones
Figure 13.2

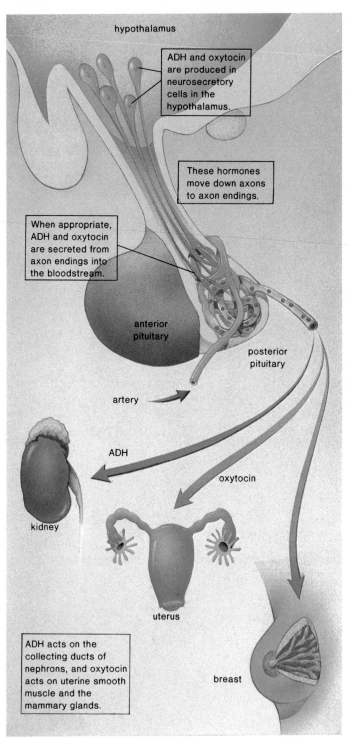

hypothalamus

ADH and oxytocin are produced in neurosecretory cells in the hypothalamus.

These hormones move down axons to axon endings.

When appropriate, ADH and oxytocin are secreted from axon endings into the bloodstream.

anterior pituitary

posterior pituitary

artery

ADH

oxytocin

kidney

uterus

breast

ADH acts on the collecting ducts of nephrons, and oxytocin acts on uterine smooth muscle and the mammary glands.

Hypothalamus
Figure 13.3

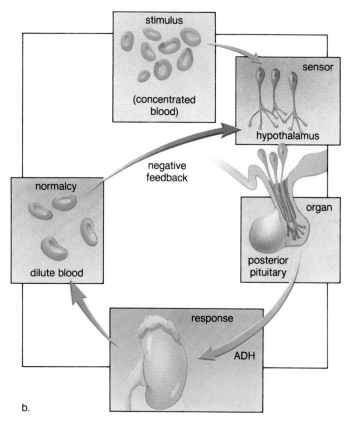

stimulus

(concentrated blood)

sensor

hypothalamus

negative feedback

normalcy

organ

dilute blood

posterior pituitary

response

ADH

b.

Regulation of ADH Secretion
Figure 13.4

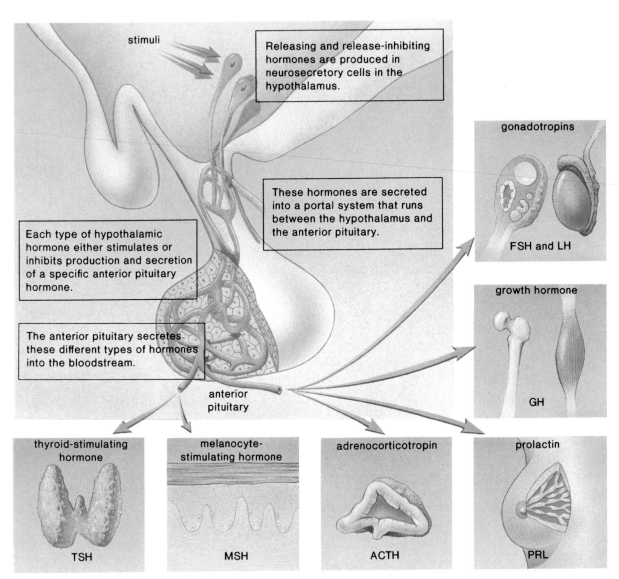

stimuli

Releasing and release-inhibiting hormones are produced in neurosecretory cells in the hypothalamus.

These hormones are secreted into a portal system that runs between the hypothalamus and the anterior pituitary.

Each type of hypothalamic hormone either stimulates or inhibits production and secretion of a specific anterior pituitary hormone.

The anterior pituitary secretes these different types of hormones into the bloodstream.

anterior pituitary

gonadotropins

FSH and LH

growth hormone

GH

thyroid-stimulating hormone

TSH

melanocyte-stimulating hormone

MSH

adrenocorticotropin

ACTH

prolactin

PRL

Hypothalamus and Anterior Pituitary
Figure 13.5

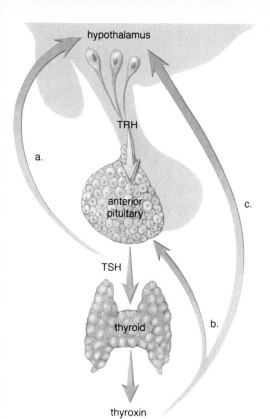

hypothalamus

**Hypothalamus-Pituitary-Thyroid
Control Relationship**
Figure 13.8

TRH

a.

anterior
pituitary

c.

TSH

thyroid

b.

thyroxin

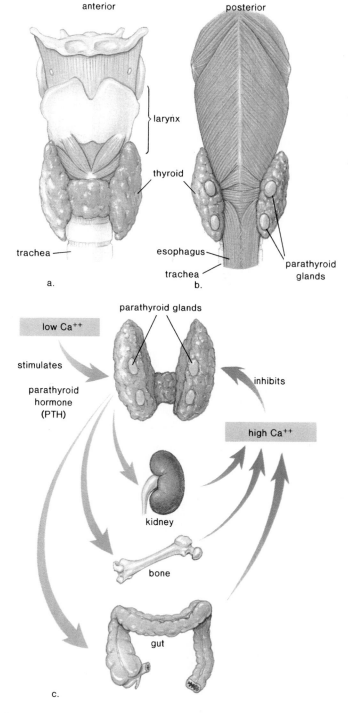

anterior

posterior

larynx

thyroid

trachea

esophagus

trachea

parathyroid
glands

a.

b.

parathyroid glands

low Ca++

stimulates

inhibits

parathyroid
hormone
(PTH)

high Ca++

kidney

bone

gut

Thyroid and Parathyroid Glands
Figure 13.13

c.

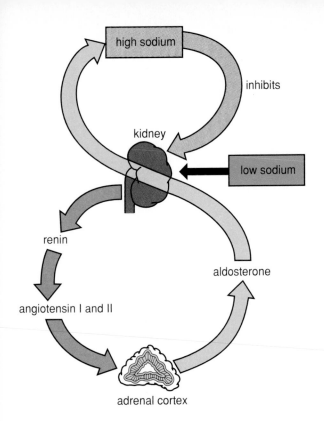

Renin-Angiotensin-Aldosterone System
Figure 13.14

high sodium

inhibits

kidney

low sodium

renin

aldosterone

angiotensin I and II

adrenal cortex

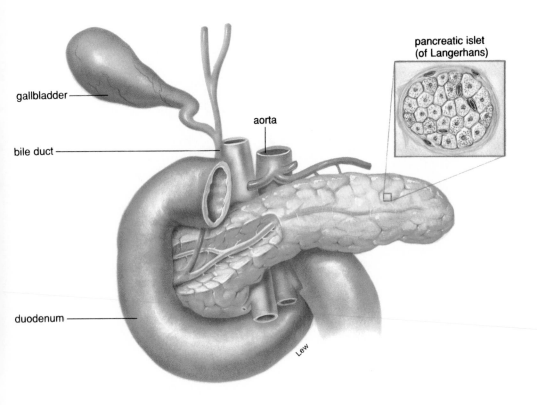

gallbladder

bile duct

aorta

duodenum

pancreatic islet
(of Langerhans)

Lew

Pancreas Anatomy
Figure 13.17

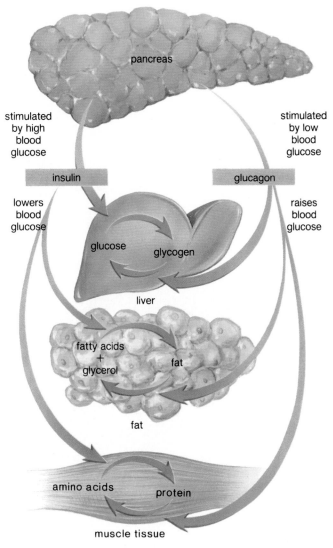

Insulin and Glucagon Functions
Figure 13.18

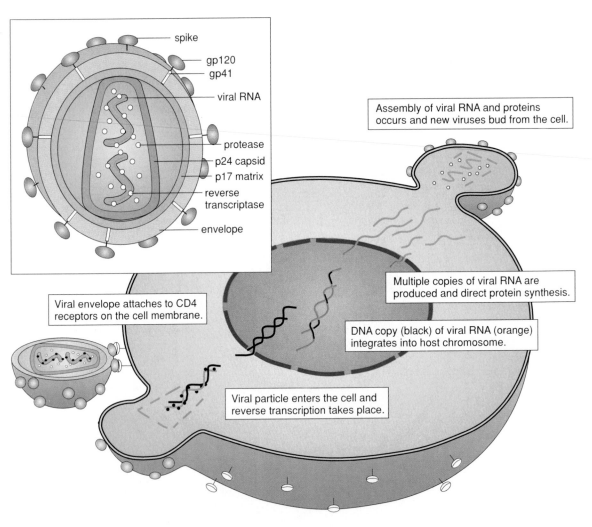

spike

gp120

gp41

viral RNA

protease

p24 capsid

p17 matrix

reverse transcriptase

envelope

Assembly of viral RNA and proteins occurs and new viruses bud from the cell.

Viral envelope attaches to CD4 receptors on the cell membrane.

Multiple copies of viral RNA are produced and direct protein synthesis.

DNA copy (black) of viral RNA (orange) integrates into host chromosome.

Viral particle enters the cell and reverse transcription takes place.

Anatomy and Reproductive Cycle of HIV
Figure A.5

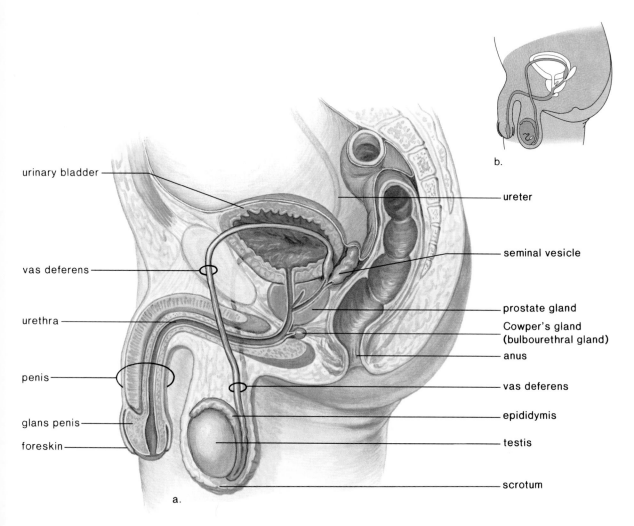

urinary bladder

vas deferens

urethra

penis

glans penis

foreskin

a.

b.

ureter

seminal vesicle

prostate gland

Cowper's gland
(bulbourethral gland)

anus

vas deferens

epididymis

testis

scrotum

Male Reproductive System
Figure 14.1

Testis and Sperm Anatomy
Figure 14.2

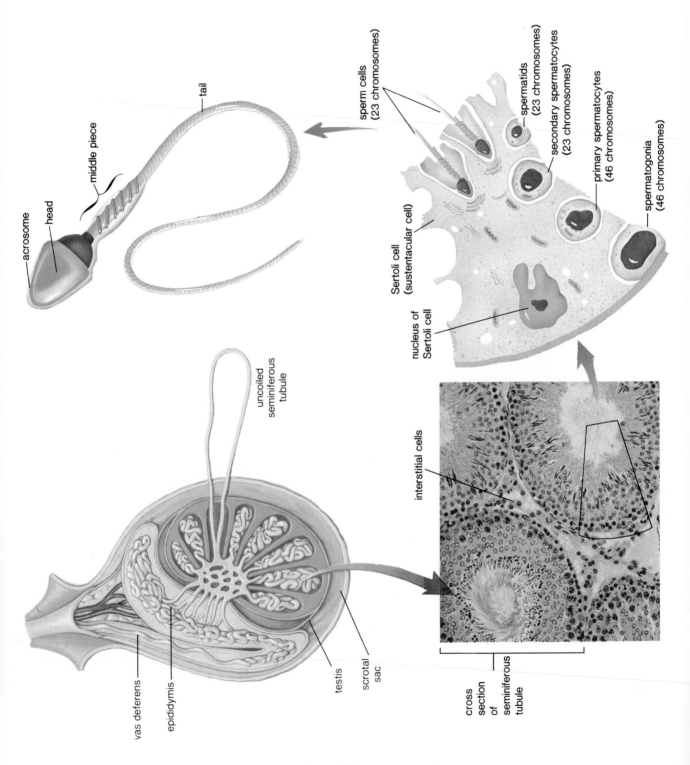

acrosome

head

middle piece

tail

sperm cells
(23 chromosomes)

spermatids
(23 chromosomes)

secondary spermatocytes
(23 chromosomes)

primary spermatocytes
(46 chromosomes)

spermatogonia
(46 chromosomes)

Sertoli cell
(sustentacular cell)

nucleus of
Sertoli cell

uncoiled
seminiferous
tubule

vas deferens

epididymis

testis

scrotal
sac

interstitial cells

cross
section
of
seminiferous
tubule

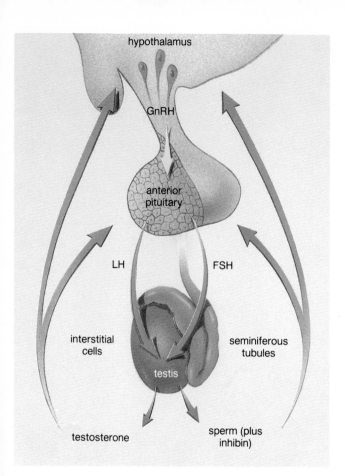

**Hypothalamus-Pituitary-Testes
Control Relationship**
Figure 14.4

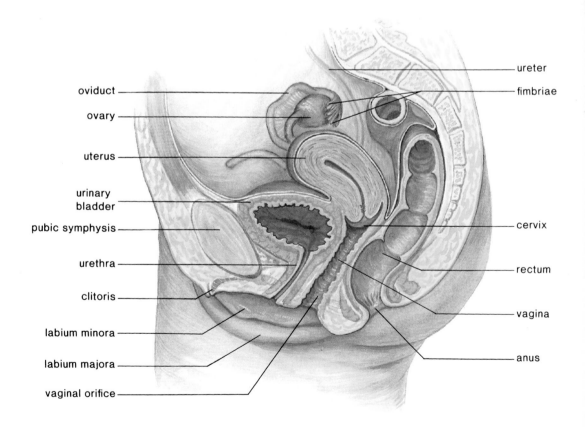

Female Reproductive System
Figure 14.5

107

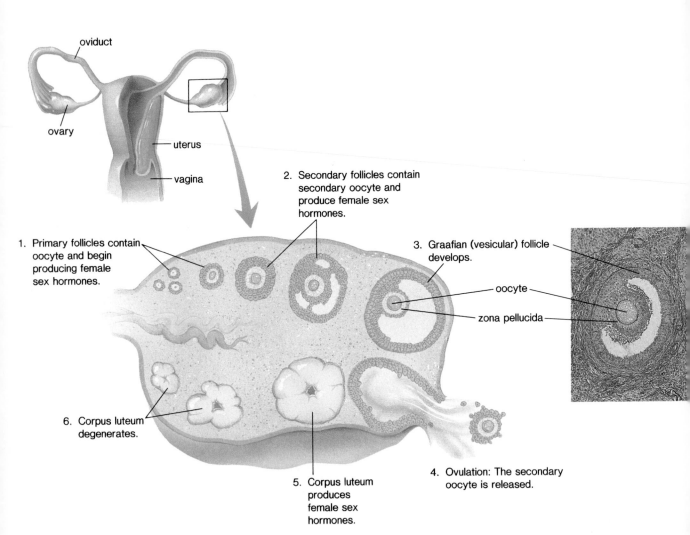

oviduct

ovary

uterus

vagina

1. Primary follicles contain oocyte and begin producing female sex hormones.

2. Secondary follicles contain secondary oocyte and produce female sex hormones.

3. Graafian (vesicular) follicle develops.

oocyte

zona pellucida

4. Ovulation: The secondary oocyte is released.

5. Corpus luteum produces female sex hormones.

6. Corpus luteum degenerates.

Anatomy of Ovary and Follicle
Figure 14.6

The hypothalamus produces GnRH (gonadotropic-releasing hormone).

GnRH stimulates the anterior pituitary to produce FSH (follicle-stimulating hormone) and LH (luteinizing hormone).

FSH stimulates the follicle to produce estrogen and LH stimulates the corpus luteum to produce progesterone.

Estrogen and progesterone affect the sex organs (e.g., uterus) and the secondary sex characteristics and exert feedback control over the hypothalamus and the anterior pituitary.

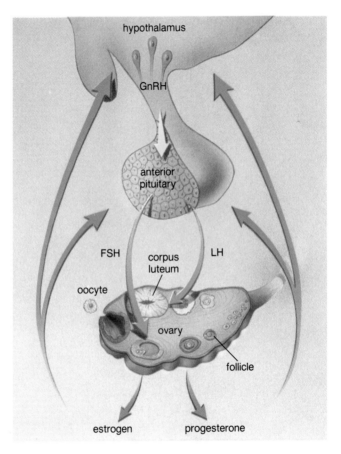

Hypothalamus-Pituitary-Ovary Control Relationship
Figure 14.8

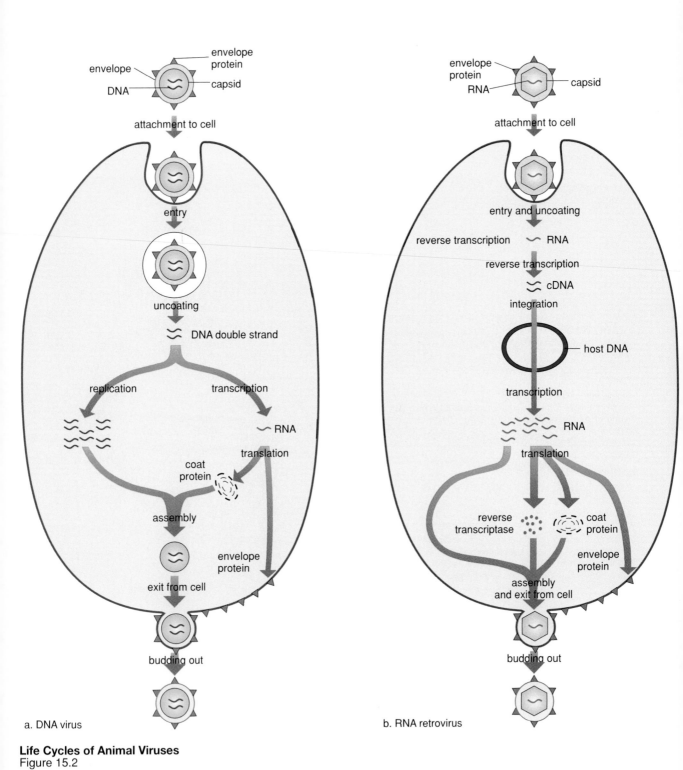

Life Cycles of Animal Viruses
Figure 15.2

a. DNA virus

b. RNA retrovirus

Reproduction in Bacteria
Figure 15.7

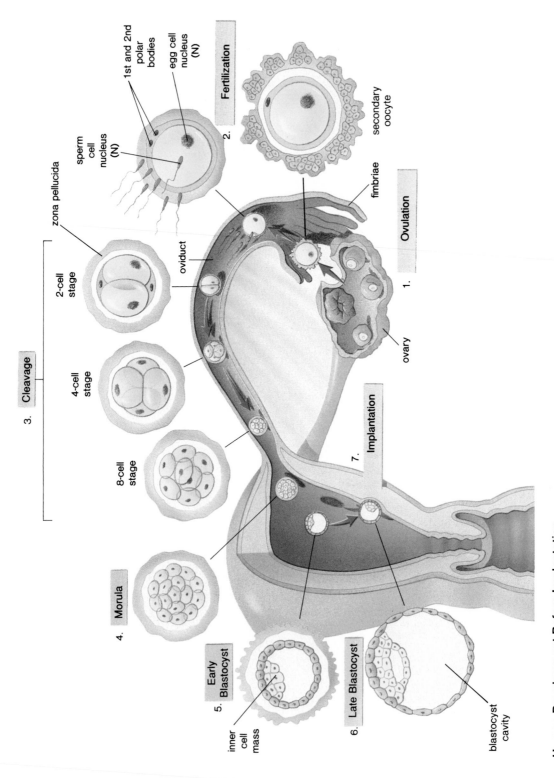

1st and 2nd polar bodies

egg cell nucleus (N)

Fertilization

sperm cell nucleus (N)

zona pellucida

secondary oocyte

fimbriae

Ovulation

oviduct

2-cell stage

Cleavage

4-cell stage

ovary

8-cell stage

Implantation

Morula

Early Blastocyst

Late Blastocyst

inner cell mass

blastocyst cavity

1.

2.

3.

4.

5.

6.

7.

Human Development Before Implantation
Figure 16.2

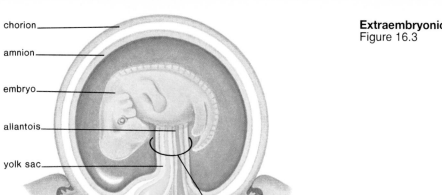

chorion

amnion

embryo

allantois

yolk sac

fetal portion
of placenta

maternal portion
of placenta

umbilical cord

chorionic
villi

Extraembryonic Membranes
Figure 16.3

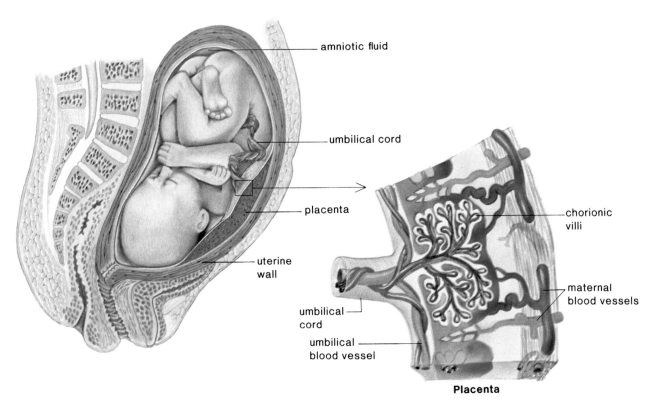

amniotic fluid

umbilical cord

placenta

uterine
wall

chorionic
villi

maternal
blood vessels

umbilical
cord

umbilical
blood vessel

Placenta

Placenta Anatomy
Figure 16.4

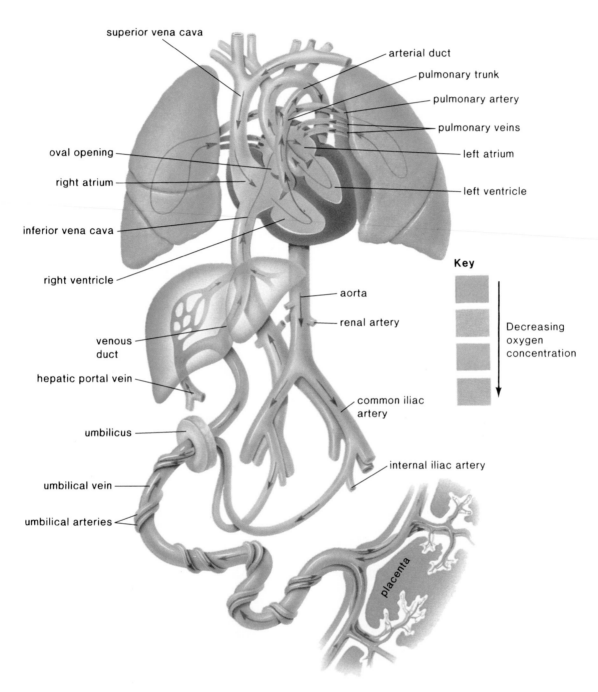

superior vena cava

arterial duct

pulmonary trunk

pulmonary artery

pulmonary veins

oval opening

left atrium

right atrium

left ventricle

inferior vena cava

right ventricle

aorta

renal artery

venous duct

hepatic portal vein

common iliac artery

umbilicus

internal iliac artery

umbilical vein

umbilical arteries

placenta

Key

Decreasing oxygen concentration

Fetal Circulation
Figure 16.6

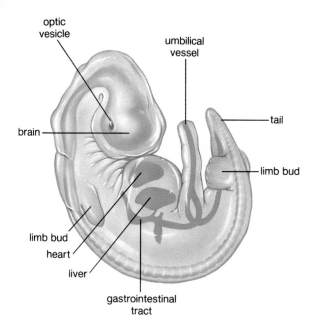

optic
vesicle

umbilical
vessel

Human Embryo at 5th Week
Figure 16.8 b

tail

brain

limb bud

limb bud

heart

liver

gastrointestinal
tract

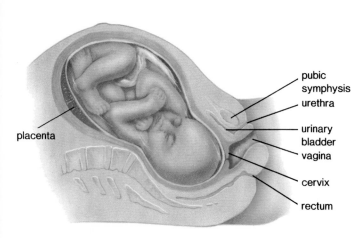

placenta

pubic
symphysis

urethra

urinary
bladder

vagina

cervix

rectum

a. 9-month-old fetus

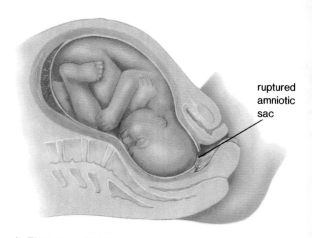

ruptured
amniotic
sac

b. First stage of birth

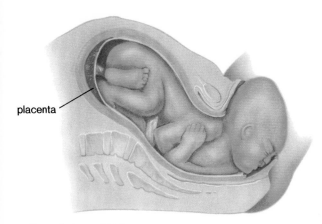

placenta

c. Second stage of birth

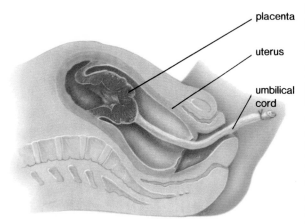

placenta

uterus

umbilical
cord

d. Third stage of birth

3 Stages of Parturition
Figure 16.13

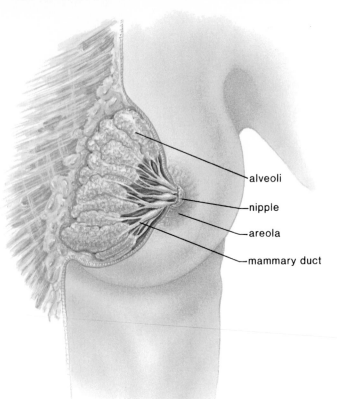

- alveoli
- nipple
- areola
- mammary duct

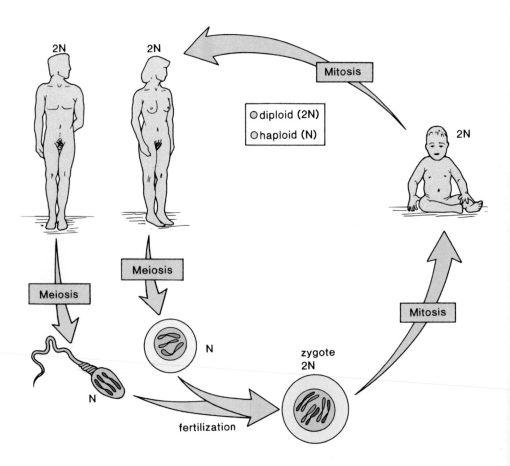

2N

2N

Mitosis

○ diploid (2N)
○ haploid (N)

2N

Meiosis

Meiosis

Mitosis

N

N

zygote
2N

fertilization

Human Life Cycle
Figure 17.6

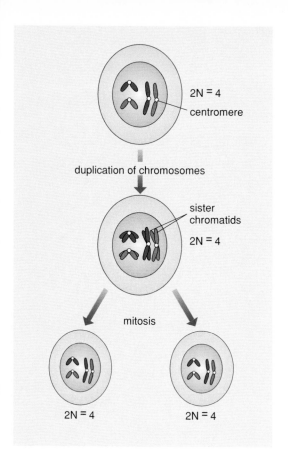

Overview of Mitosis
Figure 17.7

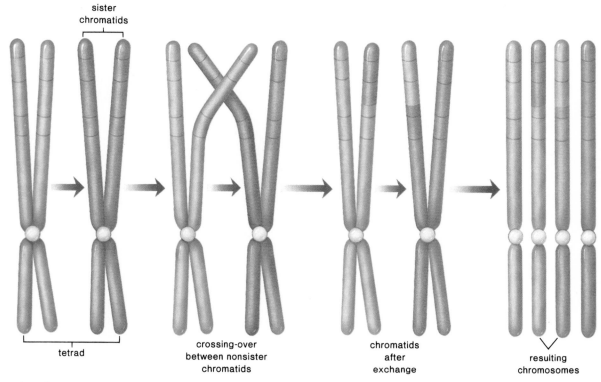

Crossing-Over
Figure 17.8

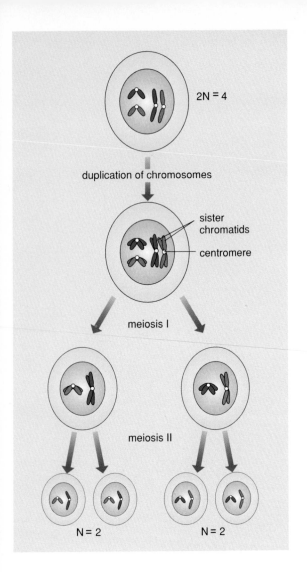

Overview of Meiosis
Figure 17.9

2N = 4

duplication of chromosomes

sister chromatids

centromere

meiosis I

meiosis II

N = 2

N = 2

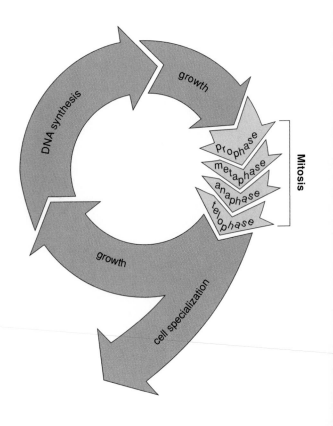

DNA synthesis

growth

prophase

metaphase

anaphase

telophase

Mitosis

growth

cell specialization

The Cell Cycle
Figure 17.10

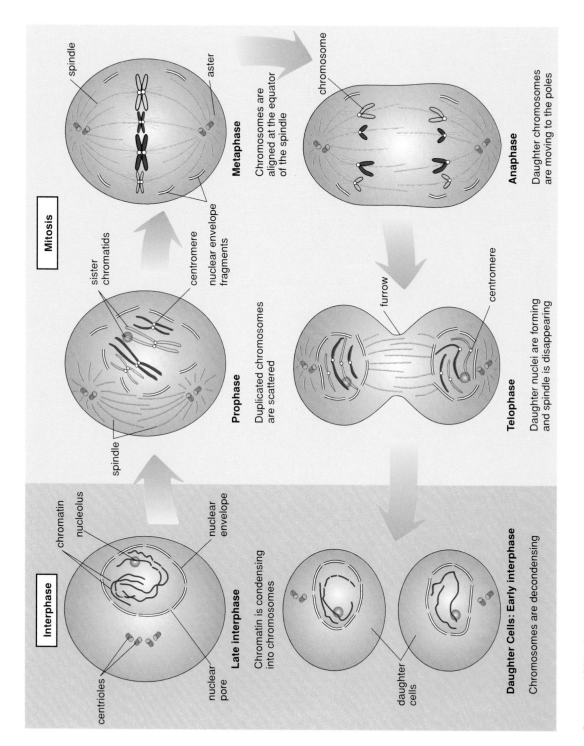

Interphase

chromatin
nucleolus
nuclear envelope

centrioles
nuclear pore

Late interphase

Chromatin is condensing into chromosomes

Mitosis

spindle

spindle

sister chromatids

centromere
nuclear envelope fragments

aster

Prophase

Duplicated chromosomes are scattered

Metaphase

Chromosomes are aligned at the equator of the spindle

chromosome

Anaphase

Daughter chromosomes are moving to the poles

furrow

centromere

Telophase

Daughter nuclei are forming and spindle is disappearing

daughter cells

Daughter Cells: Early interphase

Chromosomes are decondensing

Stages of Mitosis
Figure 17.11

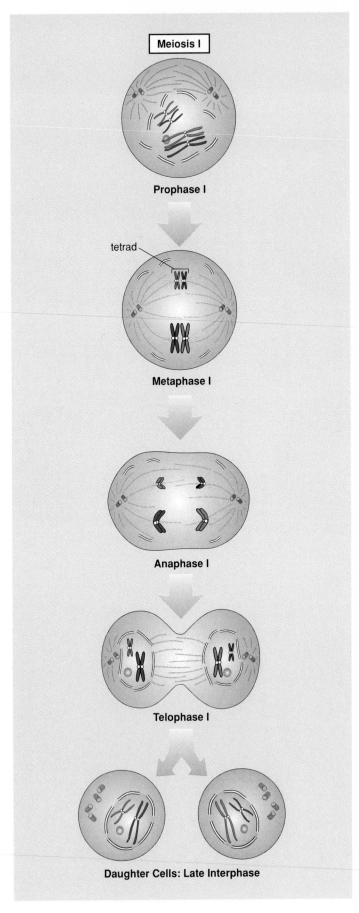

Meiosis I
Figure 17.13

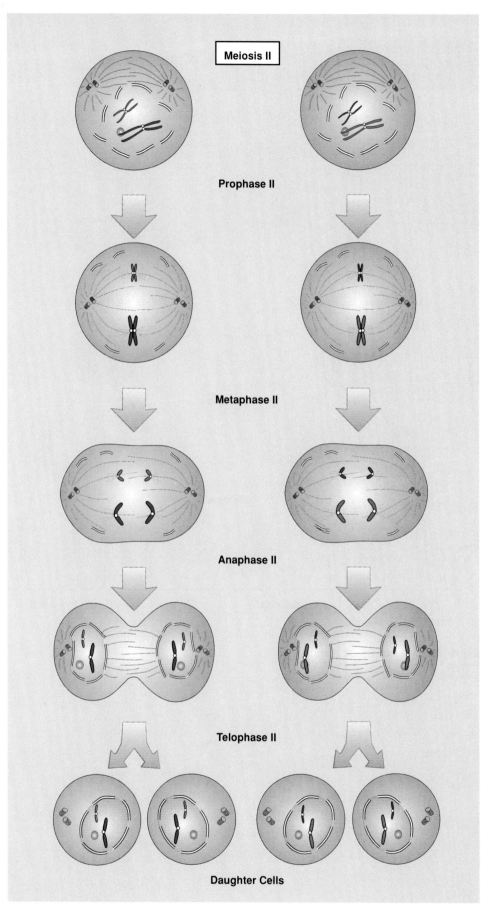

Meiosis II

Figure 17.14

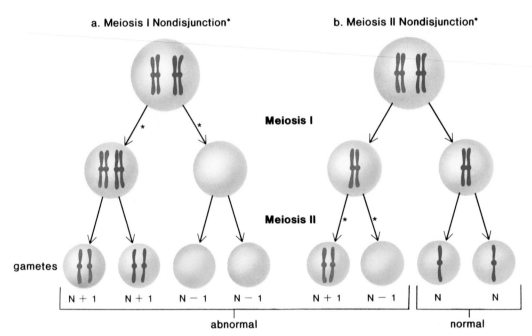

a. Meiosis I Nondisjunction*

b. Meiosis II Nondisjunction*

Meiosis I

Meiosis II

gametes

N + 1   N + 1   N − 1   N − 1   N + 1   N − 1   N   N

abnormal   normal

*point of nondisjunction

Nondisjunction of Autosomes During
Oogenesis
Figure 17.15

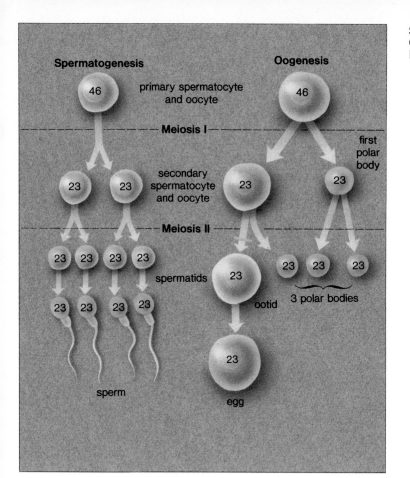

Spermatogenesis and
Oogenesis
Figure 17.16

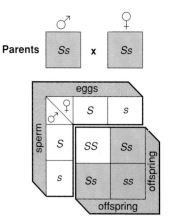

Results: 1 curly, 2 wavy, 1 straight

Genotypes   Phenotypes

□	*SS*	= Straight hair
■	*ss*	= Curly hair
▨	*Ss*	= Wavy hair

Incomplete Dominance
Figure 18.9

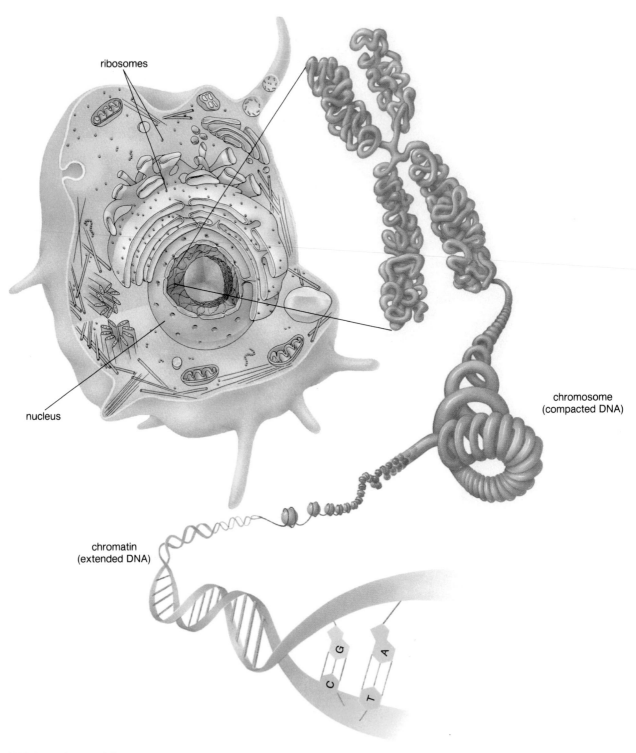

ribosomes

nucleus

chromosome
(compacted DNA)

chromatin
(extended DNA)

G

C

A

T

DNA Location and Structure
Figure 19.1

phosphate
sugar
Adenine
phosphate
sugar
Thymine
phosphate
sugar
Cytosine
phosphate
sugar
Guanine

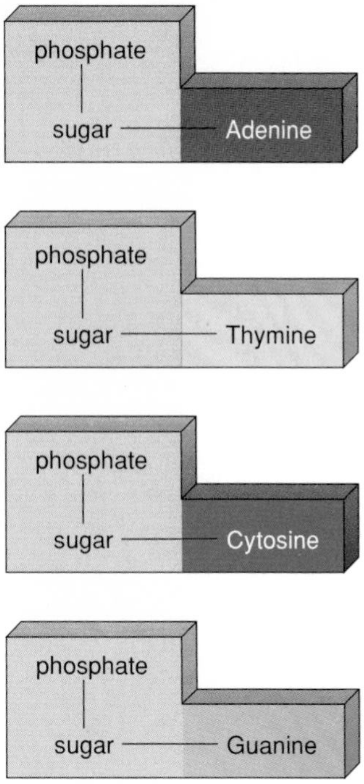

DNA Nucleotides
Figure 19.2

hydrogen bonds
P
S
A
T
S
P
P
S
T
A
S
P
P
S
C
G
S
P
P
S
G
C
S
P
P

DNA Structure
Figure 19.3

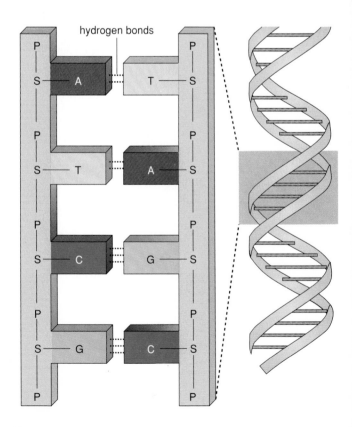

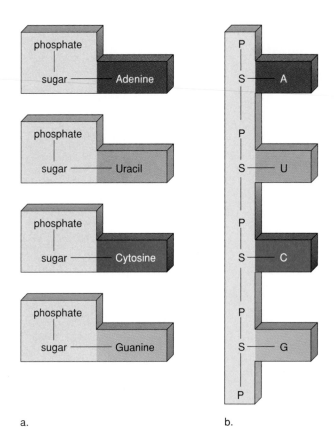

a.

b.

RNA Structure
Figure 19.4

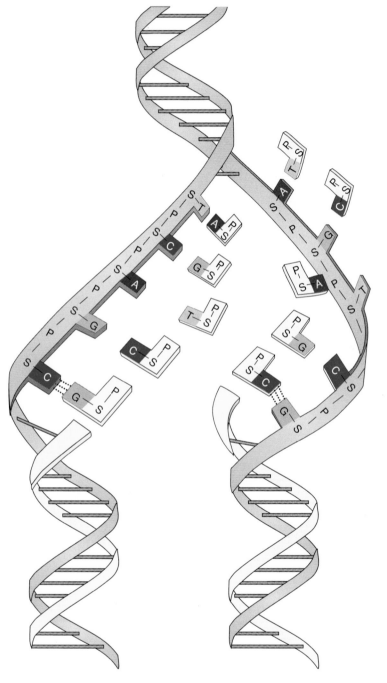

Region of parental DNA helix. (Both backbones are shown in dark color.)

Region of replication (simplified). Parental DNA helix is unwound and unzipped. New nucleotides are pairing with those in parental strands.

Region of completed replication. Each double helix is composed of an old parental strand (dark) and a new daughter strand (light).

DNA Replication
Figure 19.5

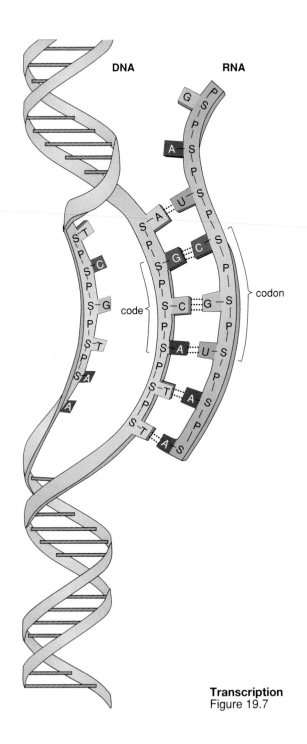

DNA

RNA

code

codon

Transcription
Figure 19.7

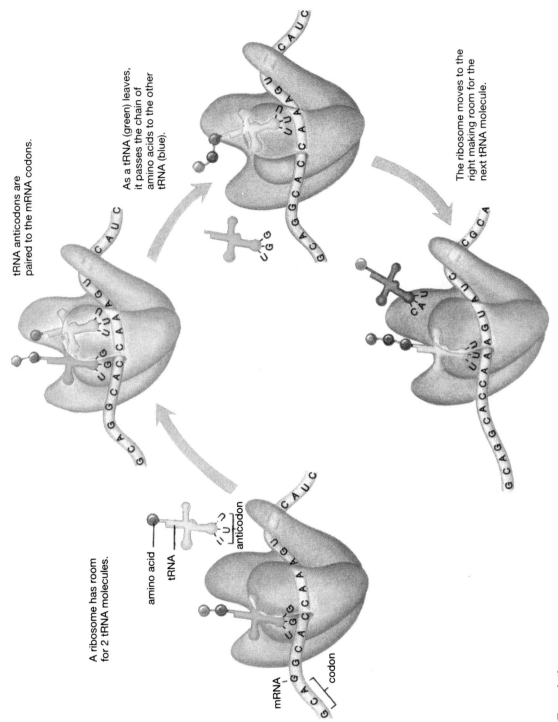

A ribosome has room
for 2 tRNA molecules.

amino acid
tRNA
anticodon

mRNA
codon

tRNA anticodons are
paired to the mRNA codons.

As a tRNA (green) leaves,
it passes the chain of
amino acids to the other
tRNA (blue).

The ribosome moves to the
right making room for the
next tRNA molecule.

Translation
Figure 19.8

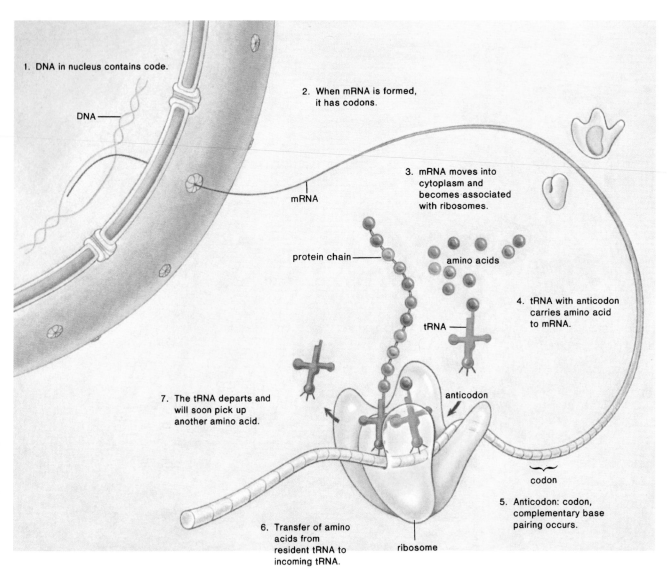

1. DNA in nucleus contains code.

DNA

2. When mRNA is formed, it has codons.

3. mRNA moves into cytoplasm and becomes associated with ribosomes.

mRNA

protein chain

amino acids

4. tRNA with anticodon carries amino acid to mRNA.

tRNA

anticodon

7. The tRNA departs and will soon pick up another amino acid.

codon

5. Anticodon: codon, complementary base pairing occurs.

6. Transfer of amino acids from resident tRNA to incoming tRNA.

ribosome

Summary of Protein Synthesis
Figure 19.9

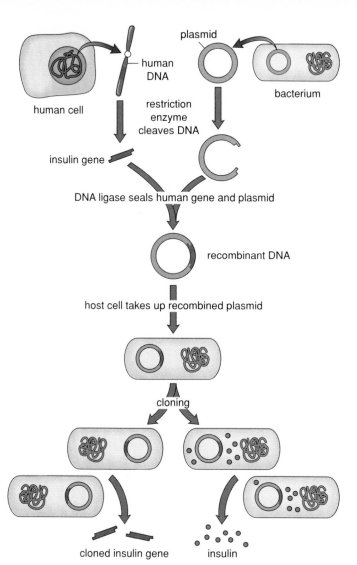

Cloning of Human Insulin Gene
Figure 19.11

plasmid

human cell

human DNA

bacterium

restriction enzyme cleaves DNA

insulin gene

DNA ligase seals human gene and plasmid

recombinant DNA

host cell takes up recombined plasmid

cloning

cloned insulin gene

insulin

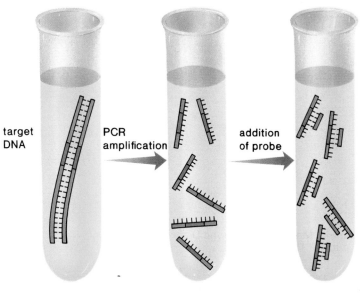

PCR Amplification and Analysis
Figure 19.13

target DNA

PCR amplification

addition of probe

a. DNA from cell

b. PCR product

c. Use of probe

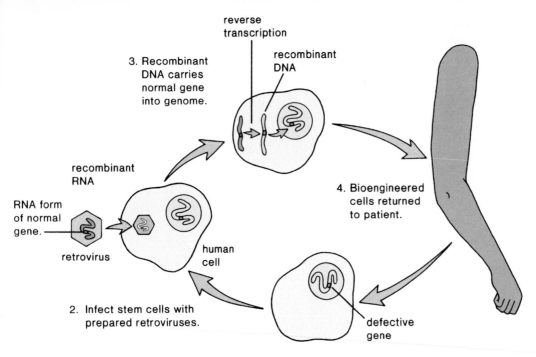

Ex Vivo Gene Therapy in Humans
Figure 19.17

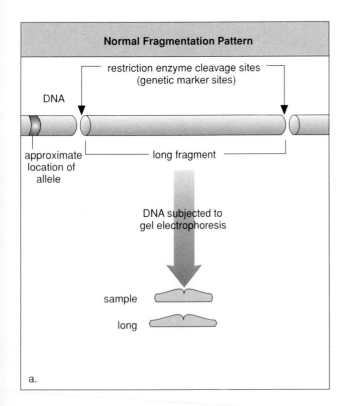

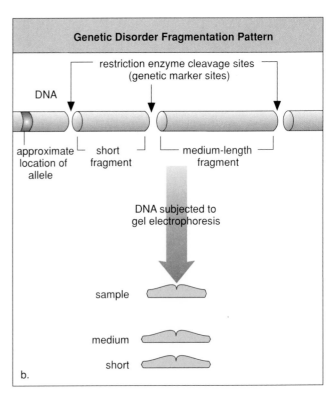

Genetic Markers for Genetic Disorder Testing
Figure 19.18

Human X Chromosome Map
Figure 19.19

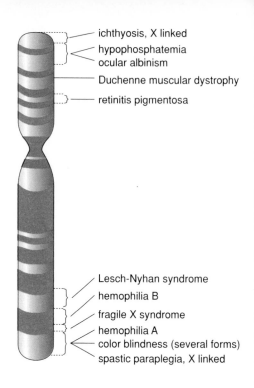

— ichthyosis, X linked
— hypophosphatemia
— ocular albinism
— Duchenne muscular dystrophy
— retinitis pigmentosa

— Lesch-Nyhan syndrome
— hemophilia B
— fragile X syndrome
— hemophilia A
— color blindness (several forms)
— spastic paraplegia, X linked

Normal Cells

Controlled growth

Contact inhibition

One organized layer

Differentiated cells

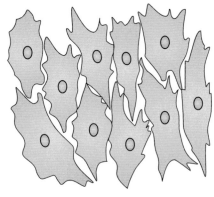

Cancer Cells

Uncontrolled growth

No contact inhibition

Disorganized,
multilayered

Nondifferentiated cells

Abnormal nuclei

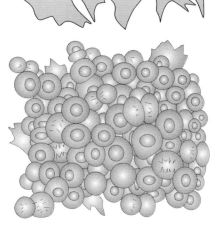

Cancer Cells Compared to Normal Cells
Figure 20.1

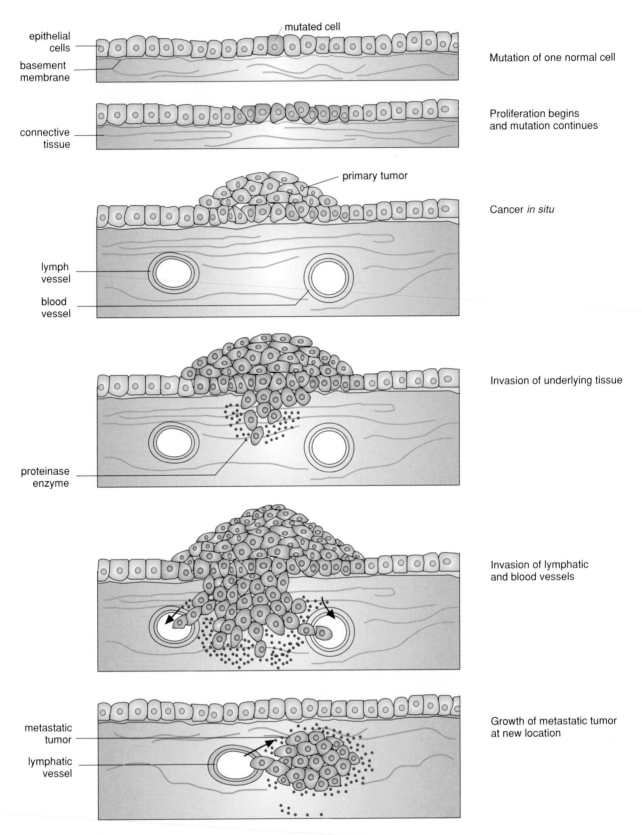

epithelial
cells

mutated cell

basement
membrane

Mutation of one normal cell

connective
tissue

Proliferation begins
and mutation continues

primary tumor

Cancer *in situ*

lymph
vessel

blood
vessel

Invasion of underlying tissue

proteinase
enzyme

Invasion of lymphatic
and blood vessels

metastatic
tumor

Growth of metastatic tumor
at new location

lymphatic
vessel

Carcinogenesis
Figure 20.2

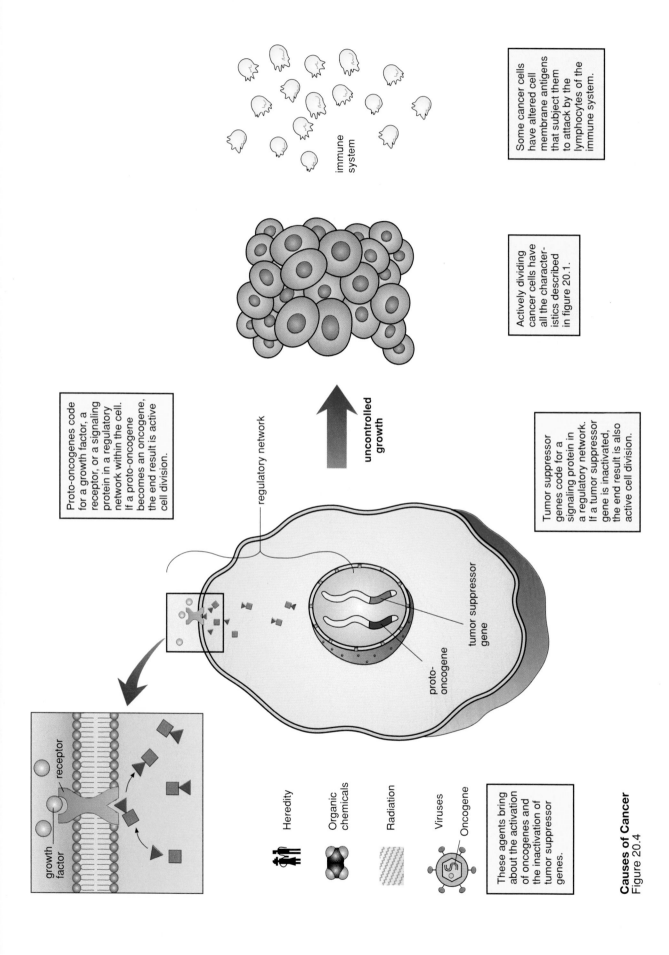

Some cancer cells have altered cell membrane antigens that subject them to attack by the lymphocytes of the immune system.

immune system

Actively dividing cancer cells have all the characteristics described in figure 20.1.

Proto-oncogenes code for a growth factor, a receptor, or a signaling protein in a regulatory network within the cell. If a proto-oncogene becomes an oncogene, the end result is active cell division.

regulatory network

uncontrolled growth

Tumor suppressor genes code for a signaling protein in a regulatory network. If a tumor suppressor gene is inactivated, the end result is also active cell division.

tumor suppressor gene

proto-oncogene

receptor

growth factor

Heredity

Organic chemicals

Radiation

Viruses

Oncogene

These agents bring about the activation of oncogenes and the inactivation of tumor suppressor genes.

Causes of Cancer
Figure 20.4

135

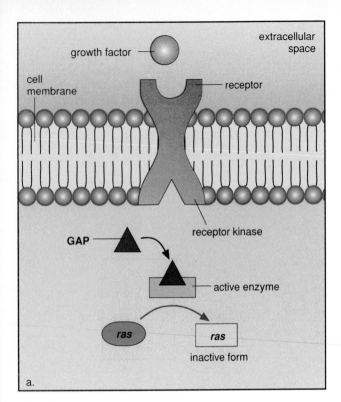
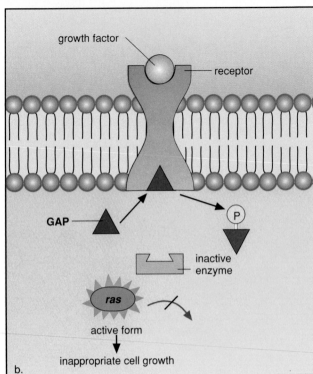

Regulatory Network Involving *Ras* Protein
Figure 20.5

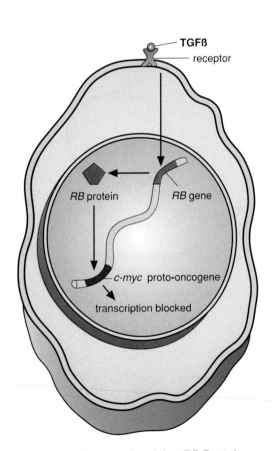

Regulatory Network Involving *RB* Protein
Figure 20.6

Normal epithelial cells

Loss of *MCC*, a tumor suppressor gene on chromosome 5

▼

Small polyp

Progression is taking place as cells divide

▼

Intermediate polyp

Activation of oncogene *ras*K on chromosome 12

▼

Large polyp

Loss of *DCC*, a tumor suppressor gene on chromosome 18

▼

Cancer *in situ*

Loss of *p53*, a tumor suppressor gene on chromosome 17

▼

Metastatic tumor

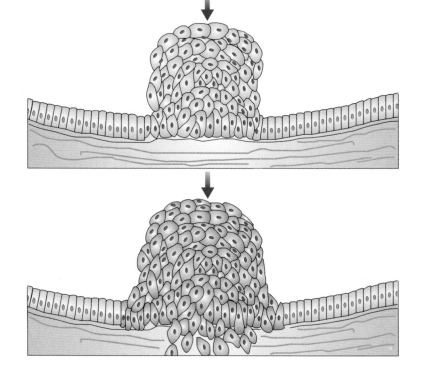

mutated cell

Development of Colorectal Cancer
Figure 20.7

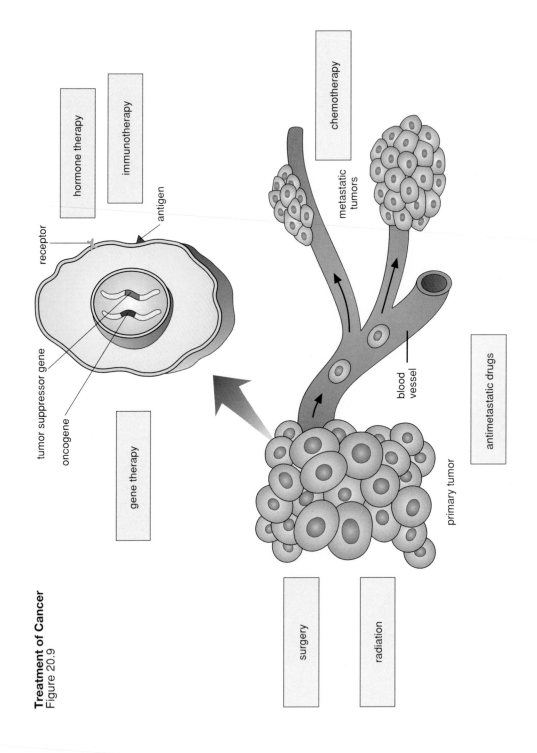

Treatment of Cancer
Figure 20.9

hormone therapy

immunotherapy

receptor

antigen

tumor suppressor gene

oncogene

gene therapy

chemotherapy

metastatic tumors

blood vessel

primary tumor

antimetastatic drugs

surgery

radiation

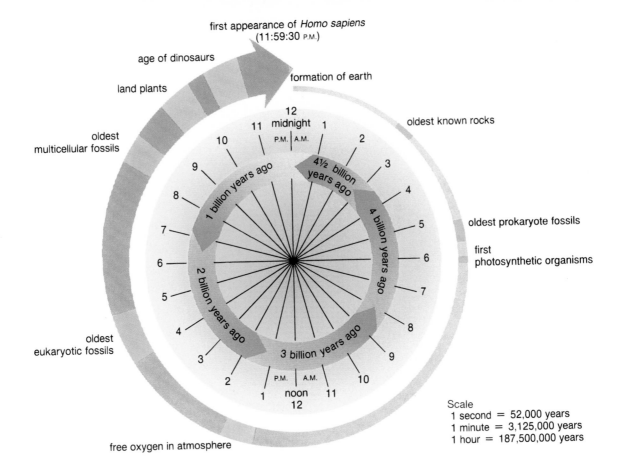

first appearance of *Homo sapiens*
(11:59:30 P.M.)

age of dinosaurs

land plants

formation of earth

oldest
multicellular fossils

oldest known rocks

oldest prokaryote fossils

first
photosynthetic organisms

oldest
eukaryotic fossils

free oxygen in atmosphere

Scale
1 second = 52,000 years
1 minute = 3,125,000 years
1 hour = 187,500,000 years

History of Earth
Figure 21.1

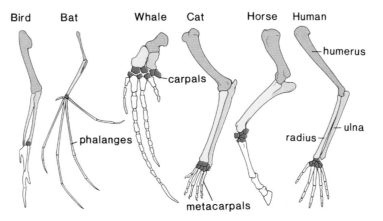

Bird   Bat   Whale   Cat   Horse   Human

humerus

carpals

phalanges

radius

ulna

metacarpals

Homologous Bones in Vertebrate Forelimbs
Figure 21.2

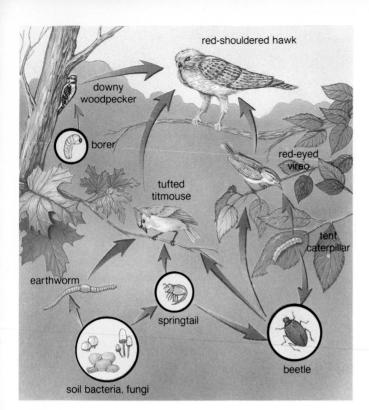

Deciduous Forest Ecosystem
Figure 22.5

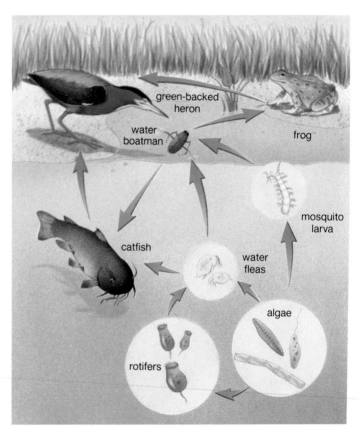

Freshwater Pond Ecosystem
Figure 22.6

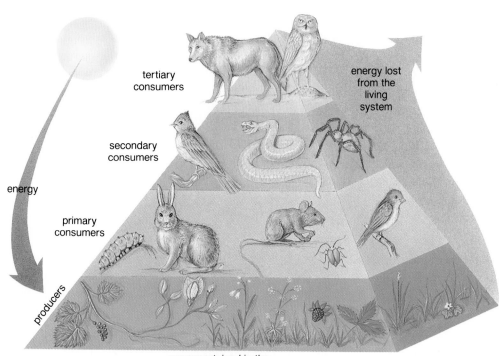

tertiary
consumers

secondary
consumers

energy

primary
consumers

producers

energy lost
from the
living
system

energy retained in the
living system

Pyramid of Energy
Figure 22.7

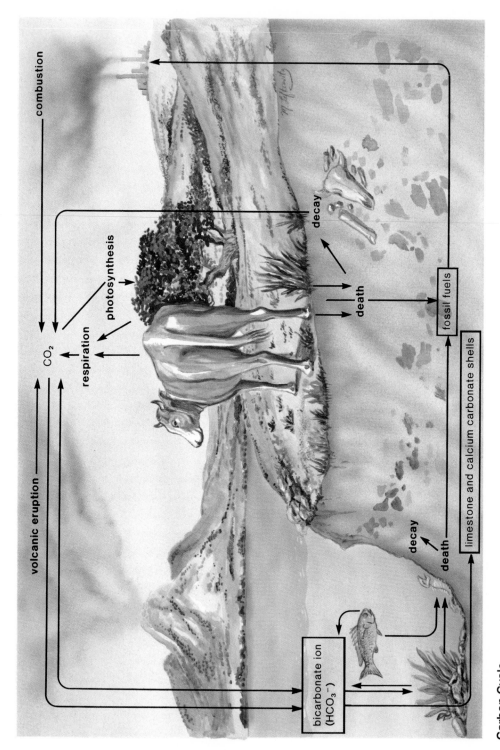

Carbon Cycle
Figure 22.9

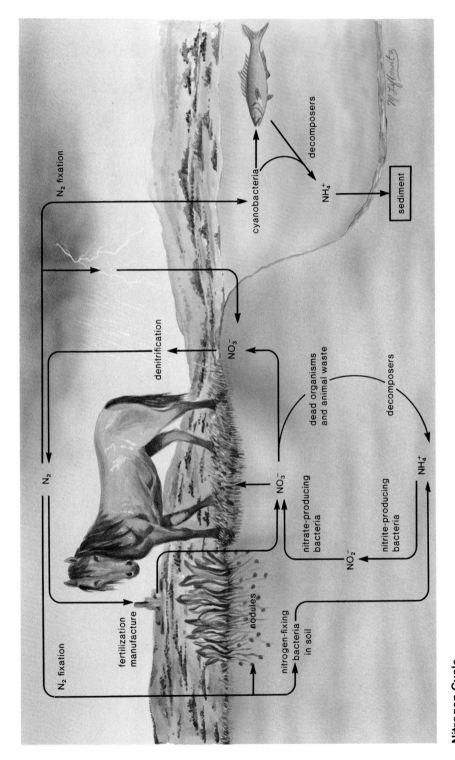

Nitrogen Cycle
Figure 22.11

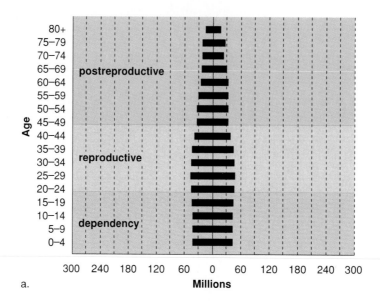

a.

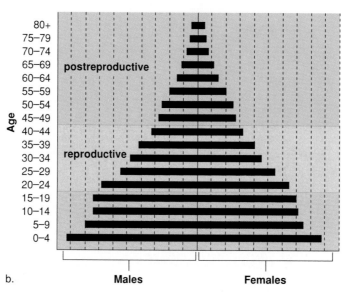

b.

Age-Structure Diagram
Figure 23.4

144

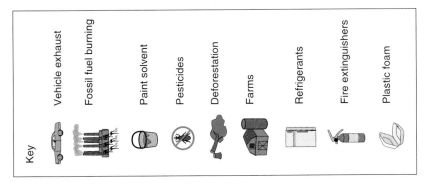

Key

Vehicle exhaust

Fossil fuel burning

Paint solvent

Pesticides

Deforestation

Farms

Refrigerants

Fire extinguishers

Plastic foam

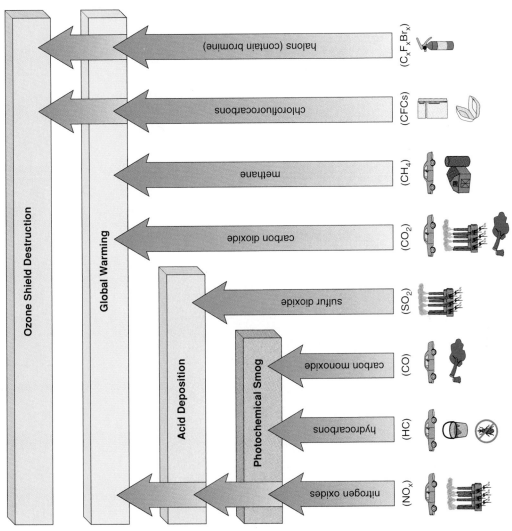

Air Pollutants
Figure 23.10

145

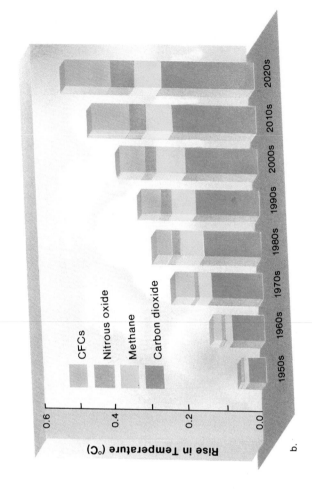

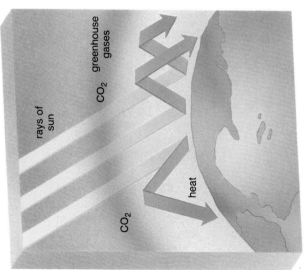

Global Warming
Figure 23.12

CREDITS

Photo

22 (2.5): © Don Fawcett/Photo Researchers, Inc.

23 (2.6): © W. Rosenberg/Iona College/BPS

25 (2.8): Courtesy of Dr. Keith Porter

31 (3.2a-d): © Ed Reschke/Peter Arnold, Inc.

32 (3.6a): © Edwin Reschke

33 (3.6b): © Edwin Reschke

34 (3.6c): © Edwin Reschke

46 (4.6c): From R. G. Kessel and R. H. Kardon, *Tissue and Organs: A Text-Atlas of Scanning Electron Miroscopy,* © 1979 by W.H. Freeman and Co.

66 (6.2): © Lennart Nilsson: HEHOLD MAN; Little Brown and Company, Boston

72a (7.1b): © John D. Cunningham/Visuals Unlimited

72b (7.1c): © Astrid and Hans Frieder Michler/SPL/Photo Researchers, Inc.

76 (7.6): From J. Rini, et al., "Structural Evidence for Induced Fit as a Mechanism for Antibody Antigen Recognition," *Science,* 255:959–965, Feb. 192. © 1992 by the AAAS.

77 (7.7): © Boehringer Ingelheim International/photo courtesy of Lennart Nilsson

80 (7.10): Courtesy of Schering-Plough. Photo by Phillip Harrington.

93 (9.6a): © J. Gennaro, Jr./Photo Researchers, Inc.

98 (10.2b): © J. D. Robertson

120 (11.12): © H. E. Huxley

129 (12.8): © Lennart Nilsson/The Incredible Machine

133 (12.14c): © Motta, Dept. of Anatomy, University "LaSppienza," Rome/SPL/Photo Researchers, Inc.

146 (14.2c): © Biophoto Associates/Photo Researchers, Inc.

149 (14.6): © Ed Reschke/Peter Arnold, Inc.

Line Art

2 (I.5): Copyright © Mark Lefkowitz.

30 (3.1) From John W. Hole, Jr., Human Anatomy and Physiology, 6th ed. Copyright © 1993 Wm. C. Brown Communications, Inc., Dubuque, Iowa. All Rights Reserved. Reprinted by permission.

35 (3.8 box) Source: Wallace, Wallechinsky, and Wallace, 1983, Significa. E. P. Dutton, Inc., NY.

36 (3.9) From Kent M. Van De Graaff, Human Anatomy, 3d edition. Copyright © 1992 Wm. C. Brown Communications, Inc., Dubuque, Iowa. All Rights Reserved. Reprinted by permission.

51 (4.13) Source: U.S. Department of Agriculture.

55 (5.3a) Copyright © Mark Lefkowitz.

56 (5.3b) Copyright © Mark Lefkowitz.

57 (5.4a) Copyright © Mark Lefkowitz.

58 (5.4b) Copyright © Mark Lefkowitz.

63 (5.9) From Stuart Ira Fox, Human Physiology, 4th ed. Copyright © 1993 Wm. C. Brown Communications, Inc., Dubuque, Iowa. All Rights Reserved. Reprinted by permission.

87 (8.10) From John W. Hole, Jr., Human Anatomy and Physiology, 5th ed. Copyright © 1990 Wm. C. Brown Communications, Inc., Dubuque, Iowa. All Rights Reserved. Reprinted by permission.

89 (9.2) From Kent M. Van De Graaff and Stuart Ira Fox, Concepts of Human Anatomy and Physiology, 3d ed. Copyright © 1992 Wm. C. Brown Communications, Inc., Dubuque, Iowa. All Rights Reserved. Reprinted by permission.

102 (10.7) Copyright © Mark Lefkowitz.

105 (10.10) Copyright © Mark Lefkowitz.

111 (11.3) From Kent M. Van De Graaff, Human Anatomy, 3d edition. Copyright © 1992 Wm. C. Brown Communications, Inc., Dubuque, Iowa. All Rights Reserved. Reprinted by permission.

116 (11.9a) From Kent M. Van De Graaff and Stuart Ira Fox, Concepts of Human Anatomy and Physiology, 3d ed. Copyright © 1992 Wm. C. Brown Communications, Inc., Dubuque, Iowa. All Rights Reserved. Reprinted by permission.

117 (11.9b) From Kent M. Van De Graaff and Stuart Ira Fox, Concepts of Human Anatomy and Physiology, 3d ed. Copyright © 1992 Wm. C. Brown Communications, Inc., Dubuque, Iowa. All Rights Reserved. Reprinted by permission.

142 (13.17) From Stuart Ira Fox, Human Physiology, 4th ed. Copyright © 1993 Wm. C. Brown Communications, Inc., Dubuque, Iowa. All Rights Reserved. Reprinted by permission.

145 (14.1a) From John W. Hole, Jr., Human Anatomy and Physiology, 6th ed. Copyright © 1993 Wm. C. Brown Communications, Inc., Dubuque, Iowa. All Rights Reserved. Reprinted by permission.

148 (14.5) From John W. Hole, Jr., Human Anatomy and Physiology, 6th ed. Copyright © 1993 Wm. C. Brown Communications, Inc., Dubuque, Iowa. All Rights Reserved. Reprinted by permission.

158 (16.13) From Kent M. Van De Graaff and Stuart Ira Fox, Concepts of Human Anatomy and Physiology, 3d ed. Copyright © 1992 Wm. C. Brown Communications, Inc., Dubuque, Iowa. All Rights Reserved. Reprinted by permission.

159 (16.14) From Kent M. Van De Graaff and Stuart Ira Fox, Concepts of Human Anatomy and Physiology, 3d ed. Copyright © 1992 Wm. C. Brown Communications, Inc., Dubuque, Iowa. All Rights Reserved. Reprinted by permission.

168 (17.15) From Robert Weaver and Philip Hedrick, Genetics, 2d ed. Copyright © 1992 Wm. C. Brown Communications, Inc., Dubuque, Iowa. All Rights Reserved. Reprinted by permission.

198 (23.4) Source: Data from World Population Profile: 1989, WP-89.